Health Intervention in
Less Developed Nations

BIOSOCIAL SOCIETY SERIES

Series editor: G. A. Harrison

1. FAMINE
Edited by G. A. Harrison

2. BIOLOGICAL ASPECTS OF SOCIAL CLASS
Edited by C. G. N. Mascie-Taylor

3. MATING AND MARRIAGE
Edited by V. Reynolds and J. M. Kellett

4. SOCIAL AND BIOLOGICAL ASPECTS OF ETHNICITY
Edited by Malcolm Chapman

5. THE ANTHROPOLOGY OF DISEASE
Edited by C. G. N. Mascie-Taylor

6. HUMAN ADAPTATION
Edited by G. A. Harrison

7. HEALTH INTERVENTION IN LESS DEVELOPED NATIONS
Edited by Stanley J. Ulijaszek

Health Intervention in Less Developed Nations

EDITED BY

Stanley J. Ulijaszek
Department of Biological Anthropology
University of Cambridge

OXFORD NEW YORK TOKYO
OXFORD UNIVERSITY PRESS
1995

Oxford University Press, Walton Street, Oxford OX2 6DP
Oxford New York
Athens Auckland Bangkok Bombay
Calcutta Cape Town Dar es Salaam Delhi
Florence Hong Kong Istanbul Karachi
Kuala Lumpur Madras Madrid Melbourne
Mexico City Nairobi Paris Singapore
Taipei Tokyo Toronto
and associated companies in
Berlin Ibadan

Oxford is a trade mark of Oxford University Press

Published in the United States
by Oxford University Press Inc., New York

A catalogue record for this book is available from the British Library

Library of Congress Cataloging in Publication Data

Health intervention in less developed nation/
edited by Stanley J. Ulijaszek.–1st ed.
——(Biosocial Society series; 7)
Includes bibliograhical refernces and index.
1. Public health–Developing countries 2. Public health Developing countries–International Cooperation. I. Ulijaszek, Stanely J. II. Series.
RA441.5.H53 1995 352. 1'09172'4–dc20 95–8024
ISBN 0 19 852302 5

Typeset by Pure Tech Corporation, Pondicherry, India

Printed and bound in Great Britain by
Biddles Ltd, Guildford and King's Lynn

ACKNOWLEDGEMENTS

We acknowledge gratefully a grant from the Parkes Foundation which enabled the proceedings of the Biosocial Society Workshop to be published.

CONTENTS

CONTRIBUTORS

Morag Bell
Department of Geography, Loughborough University, Leicestershire, UK

Richard Franceys

Water, Engineering and Development Centre, Department of Civil Engineering, Loughborough University, Leicestershire, UK

Catherine Geissler
Department of Nutrition, King's College, University of London, Campden Hill Road, Kensington, London, UK

Mary Liao
Water, Engineering and Development Centre, and Department of Geography,Department of Civil Engineering, Loughborough University, Leicestershire, UK

John Seaman
Save the Children Fund (UK), Overseas Department, 17 Grove Lane, Camberwell, London, UK

Stanley J. Ulijaszek
Department of Biological Anthropology, University of Cambridge, UK

1
Introduction

STANLEY J. ULIJASZEK

In this book, the biological and social effects of various types of health intervention on human groups and communities living in the less developed world are examined. Intervention is an integral part of development, while health is an obvious indicator of development, since it reflects improved well-being. Development is a concept which came out of the discipline of economics, and economists generally assume that economic growth results in improved human well-being. However, the associations between economic indicators and health status at national and international level are loose, at best. Unlike development, which has wide-reaching consequences, interventions are targeted changes to the biosocial environment. Like development, they are aimed at improving human well-being, health or welfare. Although such changes may be introduced in a variety of ways, intervention would not be possible in the absence of economic resources.

The concept of development includes the idea of economic improvement and the social and cultural changes which accompany it. It is linked to theories of international relations which traditionally placed societies on an evolutionary scale, with the western industrialised nations being considered most advanced, or developed. Thus, nations not yet having undergone transformations toward prosperity and economic growth are labelled undeveloped, underdeveloped, or less developed, according to their stage of economic transition. Typically, development aid for health interventions, whether in relation to water and sanitation, nutrition, or more generally, health infrastructure and service, has come from the industrialised, or more developed, nations. The stresses created by this pattern of post-colonial dependency are a recurring theme in this volume. Furthermore, the institutionalisation of aid for health interventions by way of international agencies such as the World Bank, World Health Organisation, Food and Agriculture Organisation,

United Nations Children's Fund, and a variety of non-governmental and private organisations has created a more complex web of neo-colonial dependency which continues to evolve.

Ideas about intervention, its relationship to development, international politics, health systems, public health planning, and empowerment of the poor also continue to evolve, in the light of new information, technologies, and evaluations of previous intervention strategies. In this volume, various authors examine and evaluate the effectiveness of past models of intervention and the agencies pursuing them, and consider the forces, political and otherwise, that have led to changes in approach.

Planners and health professionals have used a number of approaches in attempting to improve the general state of health in communities in developing countries, including primary health care and immunisation. Such interventions are often costly, as is the maintenance of health care structures in very poor countries. John Seaman examines the problem of meeting the recurrent costs of health service provision, under a number of different conditions. Health spending for about half of the world's population is below the World Bank minimum for supporting basic services, while in many parts of Africa, services have deteriorated since colonial times. The problem of meeting recurrent health service costs has been addressed in a number of ways, the most important of which is that packaged as 'primary health care' (PHC). This promotes the efficient use of resources within a defined health policy, by promoting better water supplies, sanitation and education, with a 'village health worker' to supply basic preventive and curative medicine, mostly financed by community support. PHC policies were adopted in most less developed countries; however, fifteen years later, the consensus view is that this has generally been a failure, although evaluation is difficult since there is no commonly held view of the measure of success. Seaman examines the primary health care initiative, discusses the problems associated with its evaluation, and considers the effectiveness of modified PHC programmes.

For an intervention to improve nutrition, it must either treat existing malnutrition, or prevent it by altering factors that are underlying causes. Nutrition interventions can take many forms, from macro-level nutrition planning at state planning office level, to nutrition education to groups of mothers in the village. They are often difficult to evaluate, but a consensus view is that undernutrition as a public health problem has declined in most parts of the developing world. Catherine Geissler describes different types of nutrition intervention, in response to the understanding of undernutrition as a multifactorial health problem. She indicates that many of the intervention approaches established from the 1940s onward are still in place in various countries; while the focus has been on protein energy malnutrition in the past, the 1990s has seen interest in the development of interventions against micronutrient deficiencies. Another focus has been on the interaction of undernutrition and infection, many nutrition programmes now combining interventions using health and specifically nutritional components. Case studies from Haiti, Thailand, the Philippines, India, and Korea are cited as examples of different types of nutrition intervention.

At the United Nations conference in Mar del Plata (1977) which launched the International Drinking Water Supply and Sanitation Decade it was concluded that 'all peoples, whatever their stage of development and their social and economic conditions, have the right to have access to drinking water in quantities and of a quality equal to their basic needs'. Implicit in this statement was the knowledge that a shortfall in either quantity or quality could lead to serious impairment of human function and quality of life. Interventions which improve either water supply or sanitation are currently believed to be the most important in attempts to reduce morbidity and mortality rates from diarrhoea in the developing world. Although the aims of such interventions are generally agreed, the approaches and the philosophies behind them have changed with time. The chapter by Morag Bell, Richard Franceys and Mary Liao gives a critique of the International Water Decade and its impact on health, and describes newer approaches in the provision of safe water.

Intervention may be an integral part of development, but it is rarely clear which has greater impact in health, intervention, or improved personal material standard of living. An underlying assumption made by economists and planners about health in relation to economic development is that increased material well-being will result in increased physical well-being. Stanley Ulijaszek shows that although the relationships between gross national product (GNP) and undernutrition, infant mortality rate (IMR) and life expectancy at birth are clear and curvilinear, GNP does not predict the level of any of these markers of health and well-being at low levels of economic prosperity. Indeed, the very high IMRs in the poorest countries may have little to do with absolute level of economic development, and might best be reduced through low-cost health intervention rather than improvements in personal material wealth. Case studies from Papua New Guinea are used to examine the effects of economic development, modernisation and health intervention on change in health and well-being, at both micro- and macro-level.

Studies in the Highlands and Highland fringe of Papua New Guinea (PNG) indicate improvements in the health and nutritional status of populations undergoing modernisation. However, it is not clear to what extent health improvements have been due to economic development and the process of modernisation which has accompanied it, and to what extent they could be attributed to specific health interventions such as primary health care and immunisation. Furthermore, it cannot be assumed that the results of these studies can be generalised to the larger PNG population. An examination of data at the national level suggests that the health and nutritional improvements associated with modernisation in the two closely studied populations of the Highlands and Highland fringe are not representative of the country as a whole, where nutritional status has declined, diarrhoeal incidence and malarial prevalence have increased, in the face of both increased GDP and greater spending on health care. However, increased immunisation coverage across the 1980s was possibly a consequence of increased gross domestic product, and the declining prevalence and incidence of diseases for which immunisations are available suggests that this intervention has been effective. Although the interactions between nutritional status, diarrhoeal disease and nutritional outcome are complex, it was expected that increased material well-being would have resulted in a decline in the incidence of diarrhoea and an improvement in nutritional status. This is not the case for PNG in general, suggesting that economic development *per se* has had little influence on health in PNG, except that it has allowed greater medical intervention.

The aim of the Biosocial Society Workshops is to examine a particular topic which lies in the biosocial domain, from a number of perspectives. Health intervention clearly has both biological and social consequences, and has a largely unwritten history. This book, based on the proceedings of the Society on 5 May 1993, at Oxford, outlines some of this history, as well as presenting current thinking about the implementation of health care and the provision of public health services among the less developed nations.

2
International Aid To Health Systems in the Developing Countries: The Problem of Recurrent Costs

JOHN SEAMAN

INTRODUCTION

International development aid is generally assumed to be 'pump priming'. Aid provides capital—which may be physical capital in the form of buildings or equipment, human capital through training or the provision of new technologies and techniques. In the classic aid 'model' it has been generally assumed that the additional costs which must arise from the investment of aid will be found by the country or people concerned. If for example a donor pays for the construction of a hospital, the 'recurrent costs' as they are often called—salaries, drugs and maintenance—will be found from the resources of the country concerned.

It has long been a rule of intergovernmental aid that no open-ended commitment can be given to the 'recurrent' costs of service provision. A long-term commitment creates dependency and contradicts the purpose of development aid. In more practical terms an aid agency with substantial commitments to long term recurrent support will have a declining pool of resources to distribute; and anyway it is not in the power of most aid bureaucracies, which are answerable to their national legislatures, to make fixed commitments to expenditure beyond a current financial year.

The exceptions to this rule are few and are found in the direct provision of services e.g. by a Christian mission, which may accept an open-ended time and financial commitment; in the short-term humanitarian relief involvements of western agencies; or rarely in projects and programmes which seek to improve the efficiency of some aspect of health provision and which are therefore 'self sustaining'. Very rarely indeed—the author is aware of only one case of a hospital in India—a donor both capitalises and provides an endowment to a project to cover its long-term costs.

In the 1960s, the period following the independence of many developing countries, this model was broadly plausible. At that time developing countries were poor relative to industrialised ones, but in terms of the absolute amounts of money available for health, their governments could support expenditure at a level which had some real meaning in terms of the services provided. There was also an expectation of future economic growth and development. International aid was useful in reducing the burden on local revenue, providing savings on foreign exchange and in assisting the development of a more extensive physical infrastructure and skilled workforce.

By the 1970s, following the effects of the rise in oil prices, many developing countries, particularly in Africa, became poorer, a process continued and accelerated by falls in the price of many primary products, on which most developing countries depend for income. In several countries the process was completed by internal conflict or outright civil war, which bankrupted several states and often disrupted any remaining administrative capacity. The decline in the level of national and private wealth, and often in the capacity of government to collect revenue, led to a situation where central expenditure on health fell to very low levels.

For example in Uganda and Ghana, real Government expenditure fell by 90% and 80% respectively during the 1970s and early 1980s (Smithson 1994). By the late 1980s in Uganda this had fallen again to about US$1/person/year. This sum was available largely in Uganda shillings: in a country dependent on imports and therefore hard currency to buy drugs and other material it was worth even less. In some countries, Nepal for example, absolute levels of health expenditure rose during the 1980s but in absolute terms remain at very low levels.

Although comparisons between countries can be deceptive—$1 in Asia will generally buy more health care than in Africa—in about 20 countries' central expenditure on health is now less than $2/person/year; perhaps a dozen more are spending less than $3/person/year (Smithson 1994). For roughly half of the world's population expenditure on health is now less than the $12 minimum currently proposed by the World Bank as the minimum required for a basic package of services (World Bank 1993).

In the early 1990s we reached the apparently paradoxical situation where substantial past aid expenditure—in 1990 total aid for health to Sub-Saharan Africa was about $1 billion (World Bank 1993)—and a substantial current level of international aid flows to health, are accompanied by the steady, and in some cases near total dissolution of the health services of the recipient countries concerned. In many cases people now have less access to both preventive and curative health services than since early colonial times. There are now countries, the Sudan for example, where in whole regions even the well-to-do cannot obtain even the most basic health provision.

The 'donors dilemma' which has intensified throughout the 1980s, has been how to accommodate the political and humanitarian need to continue to give assistance to the health sector within the general constraint that the recurrent costs of such provision could not be met from the aid budget, but could increasingly often not be met from local sources either. In this discussion, the term 'donor' includes government to government aid ('bilateral' aid); the European Union and United Nations (UN) ('multilateral' aid) and the Non-Governmental Organisations (NGO). In practice the first two dominate the international policy field.

The ways in which donors have accommodated, or failed to accommodate this situation have been complex. Donor policy and action result from a blend of many factors: the bilateral interests of the donor government concerned; lobby pressure on the donor (democratic governments being ultimately responsible to their own constituency rather than to the recipient); the need to support, often for reasons wholly unrelated to the effectiveness of aid, other international processes (for example the UN, where any individual donor may have slight effect on policy); the blurring of the relationship between policy and action by the bureaucratic process, much aid being administered by people with slight experience of the subject, a difficulty reinforced by the highly imperfect link between the actions which are taken and measurements of outcome, and the lack of any real donor accountability.

Lastly, and a major difficulty in the international debate throughout the 1980s, is a lack of some necessary fixed points in the debate on the appropriate use of aid, the two chief ones being the absence of a definition of what amounts to an adequate provision for health and what, given a reasonably efficient administration, would this cost? On the first of these, the only fixed points are the occasional international policy deliberations which find some degree of wider acceptance, in recent times the 1977 WHO/UNICEF Primary Health Care policy and the more recent World Bank World Development Report (1993) on health.

Without a reasonable international consensus on these points, discussion tends to the circular, some agencies holding that with increments in efficiency much more health benefit can be bought even with the tiny budgets available; others accepting the inadequacy of national budgets but arguing that the money could be found if less were spent on military and other Government expenditure, others again holding that the failure of Government is to collect revenue and that the difficulty could be overcome if more money were to be extracted from the population.

The responses to this tend to be that although developing country health systems are certainly often inefficient—with the progressive reduction of health budgets the pattern of spending cuts is typically asymmetric, leading to a situation where increasingly demoralised personnel are retained on declining salaries and drugs and other materials for work are not available at all—but that the absolute sums available are now so low that even substantial gains in efficiency would yield such small increments as to be irrelevant. Smithson (1994) points out that a 10% improvement in the efficiency of use of the health budget in Nepal would yield no more than $0.1/person in benefit; that although more money might sometimes be found for health from military and other sources this generally flies in the face of political reality; and that adequate administration and revenue collection itself costs money which is not available.

In short, the international aid system is a blunt weapon, which may simultaneously accommodate many different and conflicting objectives and policies and which unless controlled can and does translate these into action even in the same recipient country. In practice this control can come from only national receipient governments. Where a recipient has sufficient political and administrative substance to coordinate aid contributions aid can often be efficiently used to very effective purpose. For example China accepts aid only on its own terms: the donor is subordinate to a local and coherent policy and in the experience of the Save the Children Fund (SCF) (UK) aid can be used easily to good effect. Where, mostly in the poorer States, the recipient government has less or no control over donors, the problem of 'recurrent costs' is compounded by the a lack of any means by which external donors can be constrained to any single plan of action.

There is a case that in international health aid to the poorest countries, the 1980s have been a period of unprecedented policy and practical confusion, centrally because of the problem of the absolute poverty of many countries and the related problems of 'recurrent costs' and donor coordination. Although there are many problems in developing effective health systems other than money—there are comparatively well off countries with relatively inefficient or ineffective health systems—the case is that if the problem of a lack of basic secure local financial provision for health did not exist there would now be no fundamental difficulty in moving to provide for the basic health needs of most people in most countries. The matter is made more important by the fact that there is widespread international concern about the growth of population and large amounts of aid money available to be spent on family planning services. Practically, the provision of family planning is much eased where there is an established system of health provision.

This chapter describes some of the problems which have arisen in the provision of international health assistance to the poorer developing countries and some of the ways in which donors have accommodated to this. It draws heavily on an SCF study of the 'sustainability' of health services in five developing countries (Smithson 1994). The views presented are those of the author.

DONOR RESPONSES TO THE 'RECURRENT COST PROBLEM'

Donor responses to the 'recurrent cost problem' have not developed consecutively over time. Although since 1977 donor policy has been influenced by two major international policy reviews, the World Health Organisation/United Nations Children's Fund (UNICEF) Primary Health Care Report, and the World Bank 1994 World Development Report, donors have also acted independently of agreed international policy both within and between specific donor systems.

One clear departure from international policy from the early- 1980s was a distinctive approach to health assistance promoted by UNICEF, on a sufficient scale to form a competing and parallel strand of international policy in its own right. Some other donors have actively supported international policy, some systematically and some in simultaneous combination with other, sometimes competing lines of action. For instance a single donor might support primary health care at one level, for example through the NGOs, an opposed UN initiative at another, and combine this with independent support to some aspect of tertiary care at a third. The description which follows, apart from the two major international policy initiatives describes actions running on different policy tracks, often parallel to each other but in some cases running in opposite directions, and occasionally on a collision course.

Primary health care: the efficient use of resources within a defined health policy

The 1977 WHO/UNICEF Alma Ata conference on Primary Health Care (PHC) (WHO 1977) marked a watershed in international views of the most appropriate approach to health. PHC has been interpreted from many different perspectives, perhaps the most common being that PHC is a practical and economical approach to health provision in poor countries. In this interpretation PHC would reduce the health burden on poor people by better water supplies, sanitation and education; some simple remedies would be provided by the 'village health worker'; and the whole would be financed by 'community support', allowing an expansion of more efficient services without further pressure on the national revenue. By some it has been seen as a means to 'deprofessionalise' health and return this to the people; by others merely as a means to ease the difficulties of delivering specific health technologies.

All of these views are true. But at a more abstract and important level PHC may be seen as a new philosophy of health which sought to acknowledge and address the tensions and contradictions which were increasingly distorting health policies in all countries. The systematic development throughout this century of effective preventive and curative technologies, the vast expansion of knowledge about the environmental determinants of health and, at least in western countries and the westernised middle classes of many other countries, the changing attitudes of people to the very nature of good health, had created a situation where no government, rich or poor, could deliver the services demanded at any acceptable cost.

In the poorer countries the combination of the availability of increasingly effective preventive technologies including vaccines, the most important of which, against measles, had been available only as a practical mass technology since the late 1970s, and increasing popular demand for access to modern, effective but increasingly expensive curative services, had caused a crisis in policy and practice. In many countries an increasing proportion of the national health budget was devoted to clinical services and often increasingly concentrated on one or more urban hospitals. The quality of rural curative provision was often in steep decline. This was most clearly seen in terms of underpaid, unsupervised and demoralised workers and an inadequate drug supply. Furthermore, it was evident that no amount of curative provision would begin to solve the basic problems of health, which in poor countries are largely determined by economic, social and environmental conditions.

Primary health care set out both principles by which the conflict of interest between curative and preventative provision could be resolved and a practical approach for allocating resources between the two. There were two principles. The first was the PHC definition of health as a normative condition. Although the Alma Ata definition of health (to paraphrase, as 'a state of complete physical, mental and social well-being') is much derided as unrealistic, it is in contrast to the then-prevailing and now still common definition of health as the absence of diagnosable disease. The second was the clear statement of social policy in the objective of equity ('health for all').

The practice was found in the acknowledgement of the need to design systems in such a way that they actually moved to meet these objectives in the most efficient way within the resources available. To do this it was argued that services should emphasise the prevention rather than the cure of ill health and cure should be directed towards the common important diseases in the country concerned not rare or technically interesting diseases.

Lastly, it was clear from the Alma Ata declaration that financial control of services should remain with the country and people concerned. Although not explicitly stated in the Alma Ata document it was and is clear that the objectives of PHC can be met only if the people and country concerned are in control of policy, which in practical terms means the control of recurrent costs. The deep policy changes and the considerable length of time which would be needed to move towards PHC would not be possible on the basis of the unpredictable financial support and policy whims of foreign donors.

PHC policies were adopted by most countries and many made a serious effort to implement national PHC systems. Fifteen years later the received international wisdom is that the PHC experiment has been a general failure. More objectively, the conclusions remain open since there is no commonly held view of the measure of success. If the measure is taken as the contribution of PHC to a 'paradigm shift' among health professionals then it might be judged that PHC has had huge success, not least in the logic increasingly applied by the governments of the richer countries to modify their health policies in the face of apparently uncontrollable increases in demands for expenditure on curative care, much of which is devoted to the care of preventable diseases.

If the measure is the practical one of the implementation of PHC systems in poor countries then the picture is more mixed. In some, mostly smaller and more easily administered countries PHC has made considerable progress. The Gambia, for example, adopted and made serious attempts to implement a national PHC system which is substantially, in the form of village health workers, a vastly expanded access to the most basic services and technologies and the more efficient use of some resources, such as drugs, in place. Any regular traveller to Africa can have no doubt about the fundamental change in thinking and practice in many countries, about the most appropriate policy directions in health or that these were brought about by PHC. This discussion raises the following points:

(i) The failure to implement PHC can often be argued most convincingly from the perspective of a lack of 'recurrent resources'. PHC turned out, probably inevitably, to be more expensive than its architects had assumed and beyond any realistic possibility within the recurrent resources available. Although much of the additional cost was to be found from 'the community' additional cost on central revenues were found in the high absolute cost of many preventive technologies. For example even organised and comparatively wel-off countries such as Zimbabwe have long outstripped any financial possibility of providing centrally financed maintenance for village water supply; in the increased burden and expectation from people of the existing health services, as PHC systems encouraged referrals which had previously not occurred; and in the need for a regional or district 'support team' for PHC for which the recurrent salary and transport costs could often not be found. Where the absolute sums being spent on clinical services were very low, the reallocation of budgets away from hospitals towards rural provision was often politically and practically impossible and given the low absolute level of hospital budgets would probably have often conferred an insignificant net benefit. The central hospital, as for example in the 1980s with Mulago in Uganda, might have been spending more than half of the entire national health budget but even at that level of expenditure was grossly underfinanced. It was at the level where drugs were often entirely unavailable and the patients were not fed. The reallocation of resources from the hospital would have effectively meant hospital closure and the reallocation of such tiny real resources to rural health as to have had no benefit.

(ii) The international aid system often accepts only one measure of practical success for its health investment, that of the improvement of measures of health, the most generally accepted and used being a reduction in mortality. This primary objective often conflicts with the reasons why people want and governments provide health services. It is probably a fair statement that most people in most countries want health services chiefly for the cure of health problems with a low risk of mortality, for symptomatic relief, for reassurance and prognosis. There is no dispute that health services should save lives: but locally this may be seen as incidental to a series of more important functions. Internationally the saving of life and particularly, sometimes exclusively, those of children, has often become the sole reason for action and the sole measure of success. The reasons for this are clear enough, and are found ultimately in the donor's responsibility to an external constituency, which wants its money spent to 'good' effect: it may be easy to argue at village level for the importance of aspirin to relieve a malaria headache which is keeping a man away from his fields; it would be much harder to appeal for funds to do so in Washington or Geneva.

PHC was not clearly associated with decline in mortality. That this was not a failure of the PHC principle is shown by the rare cases where PHC was fully implemented, for example in refugee camps where the financial and political constraints are least. The basic PHC package of preventive and curative services was regularly associated with falls of Infant Mortality Rate (IMR) to less than 50, and in Thailand below 30, a level roughly that of the UK in the 1950s.

(iii) The philosophical basis of PHC was not fully understood or if it was, it was not accepted. A real difficulty of PHC is that it required control by government and people and that a condition of ownership is the control of resources. Agencies with a need to 'own' projects and large external funds which had to be spent in countries with PHC policies and very limited recurrent resources presented a contradiction. The nature of PHC was that it should not require large new capital inputs beyond the 'pump priming' of the new system, for example the provision of some additional vehicles and other start-up costs. For NGOs with smaller resources the problem was less—indeed the existence of widely acceptable coherent national PHC policies did a great deal to put order and direction into NGO work and to eliminate the eccentric NGO 'health project', local in effect and with a design and practice unique to that NGO. For the first time the more sophisticated NGOs could fall in wholly with a clear and acceptable government policy, pursue a careful line of recurrent cost minimisation, and feel that they were working towards a national rather than a purely local objective.

(iv) PHC required a sustained and consistent commitment of donor support over time. PHC involved a radical reorganisation of services, a process which takes decades. International thinking and funding tends to cycles which run by years.

'Vertical PHC': the UNICEF 'GOBI' initiative

The origins of this policy, which began in the early 1980s, are not wholly clear. The intellectual foundation seems to have been a single academic paper (Walsh and Warren 1980); the elements of the logic to have been that the continuation of the appalling health conditions, and the high mortality of children in the developing countries was intolerable given that simple effective and 'cheap' technologies had been developed which could substantially relieve this. In the 1970s there had been huge advances in vaccine technology and in the development of Oral Rehydration Solution, a simple effective and cheap remedy for diarrhoea. Vaccines were prepared which were more heat stable and improvements were made in the 'cold chain' (refrigeration and other aspects of vaccine management) by WHO through the 'Expanded Programme on Immunisation' The UNICEF initiative centred on 'GOBI'—growth charts, oral rehydration, breast feeding and immunisation—and its aim was to expand vastly the use of and access to these technologies.

In practice these technologies are not of equal value. A fair generalisation might be that growth charts have proved most useful in the context of relatively high education levels and services sufficiently sophisticated to intervene effectively when growth faltering is observed. Breast feeding is of course entirely desirable but still the general practice in most poor countries—the problems and risks of bottle feeding are real but are geographically relatively restricted. Operationally the real interest centred on immunisation and to a lesser extent on oral rehydration therapy.

UNICEF moved to expand immunisation programmes in the developing countries on a targeted basis, the aim being 'Universal Child Immunisation' (UCI), which was defined as 80% of all children in the World fully immunised by 1990. In the better-off developing countries progress was rapid. For example China, which had and has an expanding economy and health system, extended immunisation coverage dramatically, a process supported by the international donors on both a bilateral and multilateral basis. In the poorer countries GOBI confronted a real problem: the technological possibility was not matched by a means to deliver it.

In these countries the approach taken, known in aid jargon as a 'vertical' approach, was in essence to set up parallel systems for vaccine delivery, using the pre-existing health system to a greater or lesser extent according to its usefulness. In the proper countries with the least services this meant effectively a parallel service—with its own administration, vehicles, staff and budget—with the recurrent costs found from international donations.

The costs were substantial, relative to the health budgets of many countries. For example the cost of fully immunising a child was about

$15, rising to about $20 in the geographically more difficult areas such as Sudan, where transport is more expensive.

Not surprisingly this approach was often effective in raising levels of immunisation coverage. The difficulty was that in absolute terms such systems were very costly relative to the locally available recurrent resources. As international donor contributions would eventually end, and immunisation budgets be progressively withdrawn it was inevitable that the programmes would then collapse (Seaman and Poore 1987). Since the declaration of UCI in 1990 the pattern of immunisation in the poorer countries has been of declining budgets and declining immunisation coverage. For example since 1990 coverage in Nigeria has fallen from 70% to 20% (personal communication, Research and Evaluation Division, UNICEF New York); in Ghana from full coverage to approximately 40% (Ministry of Health, Ghana 1994); and immunisation in much of northern Sudan, outside of the small urban areas, has virtually ceased.

Attempts to solve the problem

Some efforts have been made by donors to solve the problem of underfinancing. The basic approaches to this have been through devising ways of generating additional resources for health, of improving the efficiency with which the available resources are used, or some combination of the two.

The prevailing systems for generating revenue for health care, other than that obtained by any government, have a long history in the industrialised world, and include health insurance and direct payment for services. In the developing countries similar systems are also long established, for example state-run insurance schemes for paid employees in parts of Latin America and in Pakistan. In the poorer countries the provision of private services on a purely commercial basis has been proscribed by Government. In practice, the situation is more complex, the apparent distinctions between Government and private systems being blurred. Underpaid health workers in Government service may often engage in charging patients or 'moonlighting' to supplement their incomes. Where a service finds its origin in traditional practice, fee-for-service systems are frequently found at a lower level of practice such as the traditional 'dai' midwife service in Pakistan, where the levels of charging may be discretionary. PHC, as originally envisaged sought to make a clear distinction between local services which would financed by the 'community' and those supported by the central budget.

Through the 1980s there have been a succession of attempts by donors to develop systems which would give a more secure flow of resources to health systems. The chief approaches to this have been local insurance systems, generally on a small scale and mediated through an NGO; through PHC 'community financing' systems, which as a generalisation have worked consistently and well in areas where community structure already has systems for the common collection and use of resources as in the Sahel of West Africa; and systems of formal charging, which fall roughly into two groups. 'Cost-sharing' tends to be a charge to cover part of the service cost, typically involving gate fees to health facilities or charges for drugs. The latter implies passing on the full cost to the user. The potential for such systems depends upon the extent to which the population can afford to use them, an assumption which is likely to be true in proportion to the levels of charging and the distribution of income in the population.

The practical difficulties include the need for substantial capital, for example in providing rural drug stores, additional transport, and running costs for administration; the difficulties of securely administering large amounts of cash on a routine basis; that Ministries of Finance may regard the system as merely revenue and cut the health budget in proportion; and perhaps most difficult of all, that in most locations local currency must be exchanged for hard currency to procure replacement drugs from overseas, hard currency which the Government often does not have.

As a generalisation it is probably fair to say that most such schemes have often failed to generate more than a small increment in health resources, or have needed repeated topping-up of their capital. The successes have tended to be examples of 'cost-sharing' where the absolute cost to the user is low. For example SCF have capitalised a drug scheme in Khartoum with the State Government of Khartoum which covers about 3 million people and provides low-cost generic, or non-copyright, drugs which are sold and resupplied from the income. After several years of operation this supplies drugs for the greater part of the population and continues to recycle the original capital. The success of the system relies on the relative affluence and population concentration of the capital and could probably not be successfully repeated in a rural area, where administrative costs would be prohibitive.

Actions to improve the efficiency of resource use are more difficult to assess. Much effort by donors has gone into the support of administrative reform and improvement although to uncertain effect. Since 1977 there has been considerable improvement in the efficiency of drug use, low cost non-copyright 'generic' drugs replacing expensive branded drugs in wide areas of the poor world.

Continued 'conventional' investment

Throughout the entire period since 1977 conventional international investment in health systems has continued. Although the major western donors have tended to moderate, to a greater or lesser extent, their patterns of capital investment in poorer countries, others have continued to invest in major infrastructure projects apparently regardless of, or perhaps sometimes in ignorance of the implications. For example the Japanese Government has provided a major fully equipped hospital in Katmandu, which if fully opened and functional is estimated to have running costs of the order of the entire national health budget. There are numerous examples of the supply of sophisticated equipment which cannot be run or supported from indigenous resources. For example the provision in the early 1980s to the Mulago Hospital in Uganda of an autoclave which required three-phase electricity and high pressure steam, neither of which were or are likely to be available. Such cases are merely a waste, rarely a recurrent burden. More significant to the recurrent budget is the continued investment in the construction of health facilities, typically health centres, which the government cannot afford to run. African administrators have sometimes learned to manage such actions by for example posting in staff and loaning equipment for the opening of the centre; and then discreetly closing it when donor interest has passed.

Compromises

At a different level to the preceding example, donors have engaged in a series of compromises which accept that in many situations there is no immediate solution to the problem of recurrent finance, but which accept that there is a need to devise 'fixes' which will allow aid to be used to continue to support the health systems which will in turn encourage the continued flow of aid money. The methods employed include local short-term efforts to 'prop up' critical parts of the system, for example providing 'salary supplements' to key administrators a practice which is now widespread, but may have unfortunate and divisive effects within a system, as a privileged group is created. The willingness of donors to support capital expenditure has long and routinely allowed for a reduction in the recurrent costs of external projects, as no allowance has to be made for maintenance or capital replacement: in due course when replacement is required another capital grant can be obtained.

Aid from the Scandinavian countries, in particular Danida, and UNICEF have been most straightforward in their acceptance of the 'recurrent cost' problems; in some cases they are prepared to finance directly, generally through the supply of basic drugs, the recurrent costs of health services, often for a period of a decade or more. Indeed, in some countries such as Uganda and Ethiopia the underpinning of a 'free' drug supply has been critical to the effectiveness of other donors. SCF could not in some cases have usefully contributed to rural health if these drugs had not been consistently supplied to the health systems with which SCF worked.

An extension of this approach is found in 'Programme Aid', which is general support by donors to the government budget which may become the means by which recurrent costs are found to sustain the additional running costs of 'projects' started by the same or other donors. There are strong suggestions that this circular, and superficially rather extraordinary pattern of financial support is becoming more common.

THE NEW REALISM: THE WORLD BANK AND THE WORLD DEVELOPMENT REPORT

The publication by the World Bank of a World Development Report on health (WDR) (World Bank 1994) marks the first major international statement on health since Alma Ata. The WDR arrives in the context of a world which has fundamentally changed from that of 1977 following the collapse of the Soviet Bloc, when many developing countries had socialist policies and state structures.

Indeed the fact that the major international statement on health for the 1990s comes from a financial rather than a health institution is itself a statement as to the depth of the changes which have occurred.

At the risk of oversimplification, there have been two basic shifts which are relevant to international health assistance. The first is that with the end of the cold war one major reason for giving aid (if the West did not the USSR would) is no longer an influence on aid policy. The second is that giving aid to the poorest is by-and-large not seen to be central to the national interest of the richer states. The removal of competing interests in poor countries means that rich countries are now free to impose their own political agendas on poor ones.

The central portion of Western intervention has been through the International Monetary Fund/World Bank in the form of the 'economic structural adjustment programme'. This has the objective of securing an orderly financial regime by, for example reducing inflation, introducing sensible exchange rates, reducing levels of government spending, and endeavouring to secure the conditions for investment and future growth. One feature of structural adjustment, in that a reduction of central spending implies the reduction of centrally funded services employment and subsidies, is that in the short term, the poor become poorer.

The World Development Report on health falls within this general approach. In summary, the measure of the efficiency of health expenditure is in terms of the 'technical' efficiency of resource use, that is, the efficiency with which money is spent on health. This reverses the position established at Alma Ata in 1977, broadly of 'allocative' efficiency, or the use of resources to the greatest effect within a defined social policy. Practically, the WDR envisages that countries identify 'priority health problems' using the concept of the 'Disability Adjusted Life Year' (DALY) where the number of 'DALYs' gained is the ratio of the costs of the intervention required to the health benefits obtained. Serious diseases, in young people where there is an effective means of intervention, for example measles, score more DALYs and are likely to have a high priority for expenditure; many adult conditions score rather lower (perhaps surprisingly, diseases associated with AIDS may gain a very low score indeed).

In the WDR the major concession to the problem of poverty is the calculation of the $12/person minimum requirement to provide a basic package of clinical and public health services. This advance (it is the first officially accepted estimate) is somewhat defused by the proposition that in low income countries 'Governments typically spend approximately $6/person/year on health and that total health expenditure is about $14'. Whatever the truth of this statement it masks the fact that the government figure is much less than this in many countries; and private expenditure is certainly so for the large numbers of the poor in most countries.

The main practical recommendation of the WDR is in acknowledging that the private sector may be used to lift some of the burden from government budgets to allow government resources to be targeted to the poor and in formally allowing for the role of NGOs as service providers. As at Alma Ata, a role is seen for 'cost sharing' and for savings through improved efficiency, for example through the use of standardised 'packages' of basic provision and generic drugs.

The effects of these new policies on the services of the poorest countries have yet to be seen. Logically a full implementation within existing national budgets and constraints might be expected to lead to little practical change. Very modest resources would be transferred from tertiary to primary care, so modest in some countries that no obvious improvement in rural care would be expected, although there would be a sharp reduction in access by the poor to care in some urban areas. Private health provision is already extensive in many countries and presumably will in future, as now, be limited by the effective demand and government regulation; NGOs providing services vary enormously in number but tend, relative to need, to be very few in most countries and capable of giving only patchy cover.

Why the World Bank has taken this position can only be a matter of conjecture. Privately, members of the World Bank accept that there are countries where no reorganisation of the national budget and private expenditure will reveal sufficient money to provide adequately for health. Perhaps the best explanation is that to admit as the basis of policy that there are now countries which into a foreseeable future will be unable to provide adequate services for their people would do nothing to change donor policy; that donors are *de facto* shifting to recurrent support; and that to state the financial reality might be to jeopardise aid flows. Nevertheless real changes may be anticipated in many countries in the 1990s reflecting the fundamental changes in global economic policy rather than specific proposals for health. If the WDR is seen as part of the general fabric of structural adjustment it may be anticipated, in countries which accept this, that donor commitment to structural adjustment will inevitably be followed by continued growth in the effective donor commitment to pay recurrent costs. From this should follow some improvement in donor coordination and efficiency and perhaps some improvement in health provision. For purists the shift may only be, as Smithson (1994) has put it, from 'pseudo recurrent' to 'pseudo development': but the practical result may be better. The only real answer to the donors' dilemma is for poor countries to become richer.

REFERENCES

Ministry of Health (1992). Annual Report of the Epidemiology Unit, Ministry of Health, Ghana.

Seaman, J and Poore P. (1987). Good intentions, unfortunate consequences. *Lancet*, **I**, 1334.

Smithson, P. (1994). *Health financing and sustainability. A review and analysis of five country case studies*. The Save the Children Fund (UK), London.

Walsh, J. A. and Warren K. S. (1980). Selective primary health care: an interim strategy for disease control in developing countries. *Soc. Sci. Med.*, **14C**, 145.

World Bank (1993). *World Development Report*. Oxford University Press, New York.

WHO (1978). *Primary Health Care*, Geneva.

3
Nutrition Intervention

CATHERINE GEISSLER

INTRODUCTION

The term nutrition intervention is often used in the narrow definition of feeding and education programmes. However, a broader definition covers any programme or policy that has a positive impact on

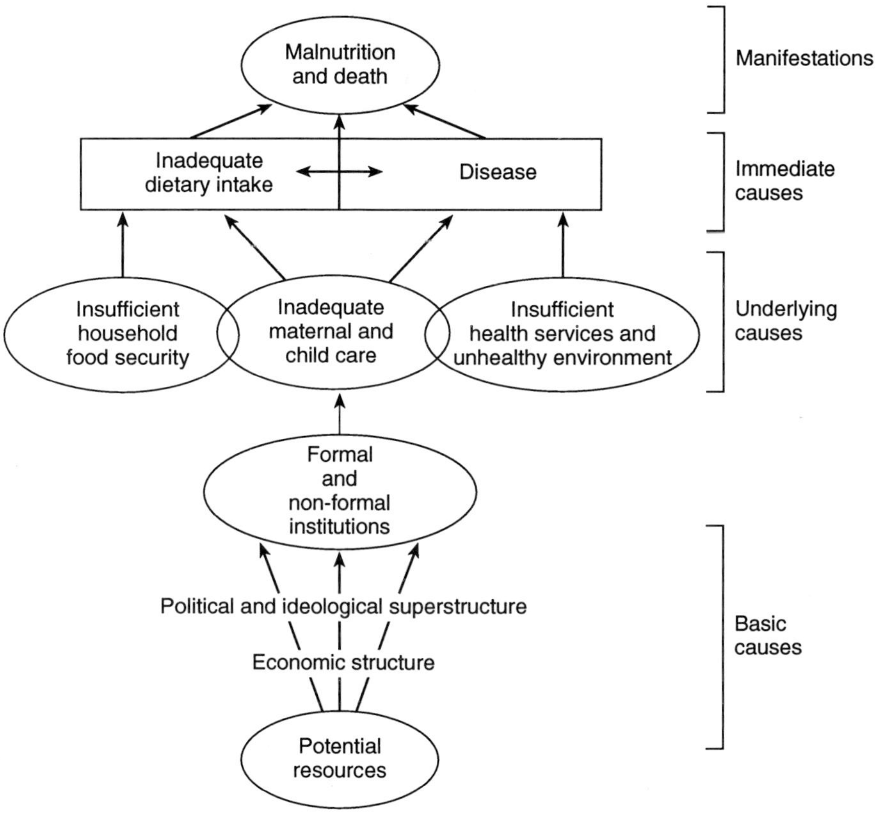

Fig. 3.1 Causes of malnutrition and death

nutritional status. It is the broader definition that is considered in this chapter. For an intervention to improve nutrition it must either treat existing malnutrition or prevent it by altering factors that are underlying causes.

TABLE 3.1. Types of nutrition improvement programme

EXPLICITLY NUTRITIONAL

Programmes related to food availability, quality, consumption and knowledge

1. *Nutrition oriented food policies*
- Agricultural production, kitchen gardens, marketing, storage, processing
- Food price and distribution control: food price subsidies; taxation; food stamps; rationing

2. *Feeding programmes*
- Mother and child: nutrition rehabilitation centres; on-site; take home
- Schools: lunch, breakfast, snack, milk
- Workers: canteens, Food for Work
- Old age: community centre; Meals on Wheels

3. *Weaning foods*
- Formulated, fermented, amylase-rich

4. *Fortification*
- Iodine, vitamin A, iron, vitamin D, vitamin C, amino acids

5. *Nutrition education*

IMPLICITLY NUTRITIONAL

. Programmes with nutritional impact through improvement of effective food demand, food utilisation and energy balance

1. *Health*
- Primary health care: immunisation, antiparasites, rehydration, basic medicines, prenatal care, health education, first aid
- Sanitation: water supply, water treatment, water storage, waste disposal, drainage and spraying, hygiene education

2. *Economic*
- Income generation
- Income maintenance: welfare benefits, unemployment benefits, child allowances, etc.
- Income substitution: e.g. subsidised basic needs

3. *Labour saving:*
- e.g. cereal mills, water storage and transport, child care creches

INTEGRATED

Combining explicit and implicit nutritional interventions: e.g. 'Applied Nutrition Programmes'; Community Development Programmes'

The causes of malnutrition are multiple and schematised in Fig. 3.1. Intervention at any level in the diagram can have a potential impact on nutrition. Therefore the types of nutrition related interventions are classified as:

1) Explicitly nutritional

These include programmes that are directly food related, including those aiming to improve food availability, accessibility, quality, consumption and knowledge.

2) Implicitly nutritional

These include programmes that have an indirect impact on nutrition through the improvement of effective food demand, food utilisation, and energy balance.

The types of programmes within these broad categories are summarised in Table 3.1.

The historical trend

There has been a charge in emphasis over the decades as knowledge of nutrition grew and governments and development agencies experienced success and failure in various approaches. This trend is outlined in Table 3.2. Feeding programmes have been part of social policy for the destitute from ancient times, during the Roman empire, in Britain initially through the Church, and more recently through secular government. Over and above these activities different approaches have been emphasised. The discovery of vitamins in the early part of this century led to the fortification of various foods with vitamins and also minerals, such as margarine with vitamins A and D, and salt with iodine.

TABLE 3.2. Evolution in emphasis of nutrition interventions

Pre-World War II: Fortification	
1940s:	Institutional feeding
1950s:	Applied Nutrition Programmes
1960s:	Green Revolution; protein supplementation
1970s:	Intersectoral Nutrition Planning
1980s:	Nutrition considerations in development programmes. Structural adjustment with a human face
1990s:	Micronutrient interventions e.g. Vitamin A, Zn

During the 1940s, the Second World War and the postwar period, there arose an emphasis on institutional feeding such as school meals, school milk, and distribution of concentrated vitamin sources to children and mothers.

In the 1950s the postwar activity of international agencies such as the Food and Agricultural Organisation (FAO) and the World Health Organisation (WHO), as well as progressive decolonisation led to a growing interest in the process of economic development and to a review of nutrition in the development process. It was recognised that individual interventions such as nutrition education were having little impact on malnutrition in developing countries and that a more integrated approach to improve the use of available resources should be encouraged. The international agencies therefore promoted what were called Applied Nutrition Programmes. These were village-based programmes addressing multiple factors of malnutrition and including such components as income generation activities, horticulture, health care, and nutrition education.

The 1960s were characterised by a more global attention to world food supply and concern that population could outstrip production. Particular fears were expressed about protein and the idea that a 'Protein Gap' existed between supply and needs. During this decade activity was centred around the production and dissemination of high yielding varieties of cereals, wheat, rice and maize, the 'Green Revolution' package, and on means to increase the production and consumption of protein.

By the 1970s a quarter century of postwar nutrition intervention experiences within the context of economic development led to further re-evaluation. It was recognised that the trickle-down theory of improved welfare with increased national wealth was not realised. It was proposed that nutrition should be a specific goal for national development on the grounds that a better nourished population will provide the human resources for effective development. To obtain this goal government policy would have to be integrated so that the underlying causes would be simultaneously addressed by the appropriate government sector such as the Ministries of Agriculture, Employment, Commerce, and Health. Intersectoral Nutrition Planning would be coordinated by a Nutrition Planning Unit in an umbrella organisation such as a Ministry of Planning or a Prime Minister's Office. This approach was fostered by development agencies such as FAO and USAID (United States Agency for International

Development) and was adopted by several countries, particularly after it was endorsed at the World Food Conference in 1974.

Much was learned from attempts to apply the rational procedures advocated by the national Nutrition Planning approach for the selection of appropriate interventions. In particular, it became clear that there was a paucity of data on which to decide nutritional priorities and the effectiveness of various interventions. In addition the feasibility of placing a priority on nutrition and of effective intersectoral coordination was cast in doubt. These hurdles led to increased research in the evaluation of interventions, the functional impact of malnutrition, and to a move from intersectoral planning to ensuring that existing sectoral interventions such as agricultural development programmes included nutrition considerations. This was the main theme of nutrition intervention in the 1980s. The other theme related to the Structural Adjustment Programmes that were being enforced by the international finance agencies, the IMF (International Monetary Fund) and the World Bank, to reduce government spending and improve balance of payments. Reduced spending resulted in cutting a variety of welfare programmes that were effective in controlling malnutrition, such as food price subsidies. The second theme of the 1980s was therefore the targeting of nutrition interventions to those most in need and the involvement of local communities in self-sustaining development programmes.

The various approaches favoured in each decade remain on-going in the 1990s in several countries and with different emphases. The theme of the current decade is yet to evolve but it appears at this early stage that interest is concentrating around micronutrient interventions, including vitamin A, vitamin E and selenium, which are being recognised as important not only for eye lesions and blindness, but also in resistance to respiratory and diarrhoeal infections, and in protection against cancers and heart disease. Another focus for supplementation, for example with zinc, is to address the widespread phenomenon in developing countries of restricted growth.

International trends in malnutrition

To what extent have these efforts to reduce malnutrition been effective on the world scale? One of the watersheds in international attention to food and nutrition was the World Food Conference in 1974, for which detailed assessments of malnutrition, in the sense of

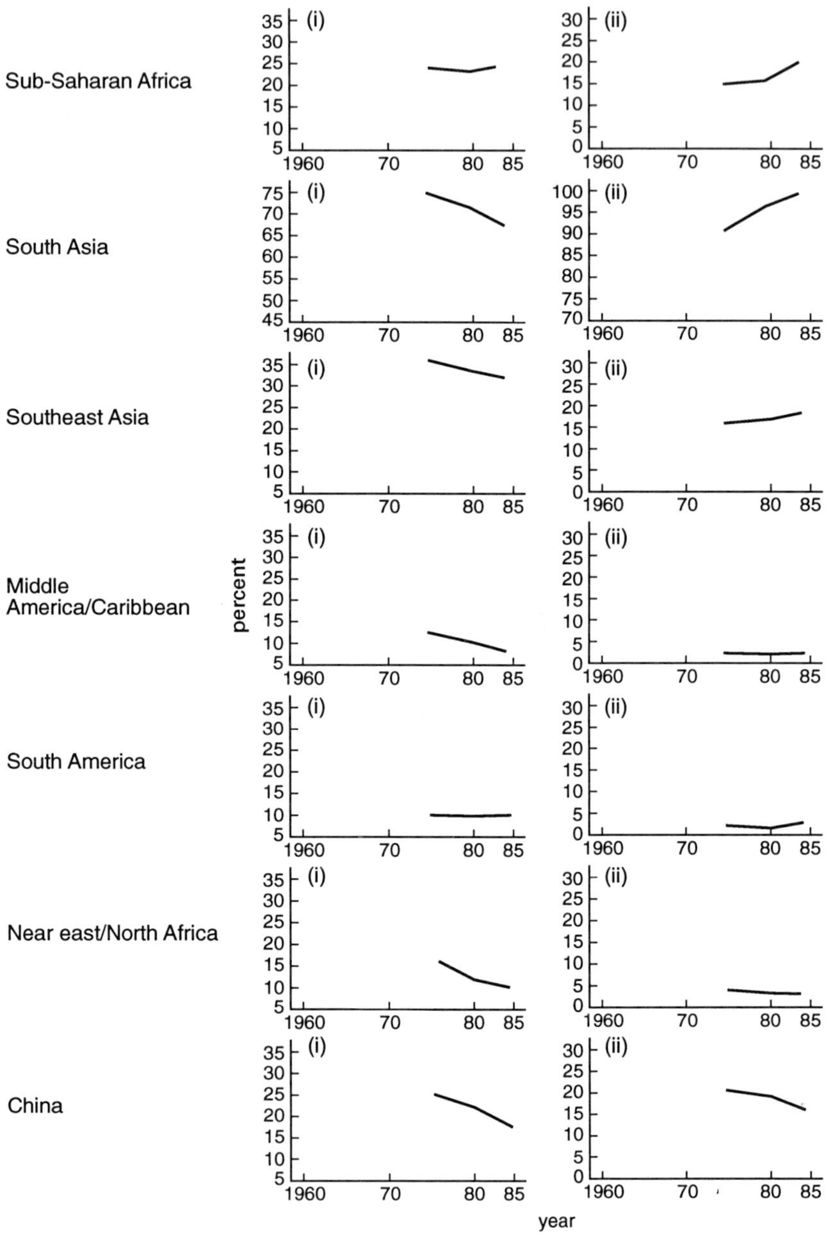

Fig. 3.2 Regional prevalences (graph (i), percentages) and numbers (graph (ii), millions) of 'underweight' children (< 2 SD weight for age). *Source*: Gillespie and Mason (1991)

undernutrition, were made. The next major international assessment was at the International Nutrition Conference in 1992. Over this twenty-year period there have been considerable changes in the percentages and numbers of underweight children in different parts of the world, as shown in Fig. 3.2. The percentage has declined in all parts of the world except Sub-Saharan Africa and South America, but the numbers have declined only in China, remained more or less the same in Central and South America and the Near East/North Africa, and increased markedly in South Asia and Sub-Saharan Africa. Most data on nutritional status relate to preschool children as these are considered the most vulnerable. However, other age groups are not immune to malnutrition.

Malnutrition of poverty and affluence

Dealing with undernutrition is the main emphasis of the development agencies, and in terms of impact on human life this is the most pressing and dramatic nutritional problem worldwide, with enormous social and economic impact. However, the nutrition related diseases commonly associated with affluence such as coronary heart disease, diabetes, obesity, dental decay and cancer afflict sections of the populations in developing countries and diseases associated with poor countries, such as anaemia, also afflict sections of the population of rich countries. Malnutrition in the sense of suboptimal nutrition is an international problem. The types of interventions that are used for undernutrition and for diseases of affluence, in both developed and developing countries overlap. In developed countries undernutrition is largely prevented by such means as adequate incomes, welfare benefits, and food fortification, which are features of specific interventions in developing countries. However, interventions to address diseases of affluence tend to be based almost entirely on education of the public as other measures such as price manipulation and rationing are considered too coercive in peacetime democratic societies.

Undernutrition is clearly related to poverty, yet some countries do better than others. This can be seen in Fig. 3.3 which shows the amount of food available for consumption in relation to energy requirements, as an indicator of food adequacy, for a large number of countries. At the same GNP some countries are in a much more favourable situation than others. Other indicators of nutrition and health such as infant mortality show similar trends. The question

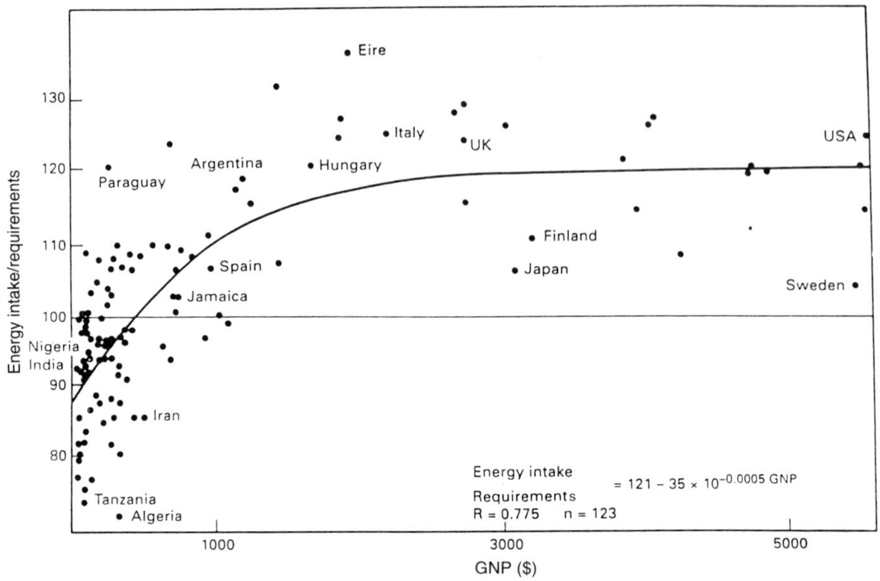

Fig. 3.3 Satisfaction of energy requirements measured against international levels of gross national product. *Source*: Geissler and Miller (1982)

then arises why some countries do better than others. What are the lessons to be learned for nutrition intervention? If a government is seriously committed to improving the basic needs of its population its needs to know how to use its limited resources most effectively. This knowledge depends on information available from the evaluation of existing programmes.

Evaluation of effectiveness

The evaluation of effectiveness of intervention is theoretically simple but practically difficult. The main aspects of nutrition that can be measured and therefore evaluated are food consumption and nutritional status. Food consumption can be assessed directly at the household level by household food inventories and at the individual level by weighed dietary intakes over several days, or by recalled or recorded diaries using household measures. At the national level it can be assessed indirectly by estimating the per capita food available for consumption using food balance sheets based on national production, imports, exports, and wastage, divided by the population. Nutritional status can be assessed by anthropometric, biochemical or clinical measures.

There is a clear need for good evaluations to ensure cost-effectiveness when budgets are tight, and to facilitate the successful transfer of programmes to other settings. The constraints to useful evaluations are the preference for evaluations of process rather than impact. When impact is measured it is usually the impact on intended beneficiaries alone that is measured, whereas the wider implications also need to be assessed: what effect does the programme have on other family members, other sections of the population, what are alternative uses of the funds, do the parameters measured assess the main impact? Evaluation tends to carry a connotation of criticism and is therefore generally unwelcomed by those implementing the programme. The cost of evaluation has to be weighed against alternative use of these funds for further action rather than study. The evaluation of small pilot projects is easier than that of wide-spread operational programmes in a multifactorial context for which a comparative control population is hard to find.

Therefore evaluation is difficult, and information tends to be partial and anecdotal but lessons have been learned in recent years about the effectiveness of nutrition interventions. Several reviews have been published, for example Berg (1987), Gillespie and Mason (1991), Dreze and Sen (1989), Gittinger *et al.* (1987), Levinson (1987),and Gwatkin *et al.* (1980) in addition to specific case studies.

TYPES OF INTERVENTIONS

The rest of the chapter presents case studies and reviews of selected specific types of intervention (Gillespie and Mason 1991) to illustrate some of the issues addressed in the introduction. These are based on nutrition interventions carried out by the main agents—governments, UN agencies and non-governmental organisations (NGOs).

Agricultural projects

The Green Revolution

The Green Revolution is an example of the prevalent erroneous assumption that increasing the supply of food will automatically reduce the extent of malnutrition. The objective of the Green Revolution was to increase the supply of staple crops by the introduction of high yielding varieties of seeds with a package of supporting inputs including irrigation, fertilisers and pesticides. The revolution started in the 1960s at the CIMMYT in Mexico (*Centro Internacional de Mejoramiento de Mais y Trigo*; International Centre for the Improvement of Maize and Barley) and at the IRRI (International Rice Research Institute) in the Philippines. It was assumed that improved techniques would increase food production and inevitably improve the economic and nutritional situation of households. The shorter cropping cycle of the improved seeds would allow double cropping and so increase both production and labour employment.

Yields and production certainly increased and market prices for staples decreased but the expected miracle was mitigated by social and economic barriers. The main success was in areas where irrigation was assured and farmers were rich enough to risk experimentation and to obtain credit. The best land was therefore used first and rapid spread was inhibited by the availability of a reliable water supply and credit for smaller farmers. The revolution has been criticised for exacerbating poverty and malnutrition amongst rural populations as the possibility of increased profits to landowners resulted in many evictions of tenant farmers, while the increased labour demand often resulted in increased mechanisation rather than employment. Where employment has increased it has been disproportionately met by poor women and the impact of this depends on the control they may have on their income and their possibilities for suitable child care. The impact on wages has varied: where there is little unemployment wages have increased; where mechanisation and unemployment predominate wages have fallen. Wages are increasingly paid in cash and this has had adverse consequences during high inflation and increasing food prices. The impact on consumers of reduced prices of staples can be beneficial. However, this has in many cases been offset by employers reducing wages. Traditional intercropping of grain crops with leguminous crops was no longer possible with mechanisation so the availability of peas and beans was reduced and their price increased, with a detrimental effect on variety of diet. Many analysts have therefore been critical of the social impact of the Green Revolution, concluding that it has merely strengthened the economic position of rural and urban elites and has been detrimental to the poor in economic and nutritional terms. However, Pinstrup-Anderson and Hazel (1985) argue that the early undesirable effects of the revolution were due to inappropriate policies in relation to land tenure and access to inputs.

The revolution increased aggregate food production but exacerbated inequalities. Rural and urban elites benefited, the poor suffered. The poor could benefit from the increased production if stocks are used in public distribution and nutritional support schemes, for example targeted food subsidies, food for work, and supplementary feeding. It has demonstrated a need to change credit policies away from a bias against the poor.

The Green Revolution is another example to support the relatively recent widespread recognition that undernutrition is a problem of inability to buy or have access to food rather than of food supply.

An irrigation project in Haiti

This example demonstrates features of the evaluation of nutritional impact (Brun *et al.* 1990). The example is a large irrigation scheme, ODPG (*Organisme de Developpement de la Basses Plaine de Gonaives*), initiated by FAO in 1972. It covered an area of 2750 hectares cultivated by families with a total population of 20 000. To evaluate the impact of the scheme a contiguous control area was selected. For rapid evaluation ten households from the ODPG area and ten from a similar ecosystem in the control zone were studied. Prior to the scheme less than a quarter of the land was seasonally irrigated with river water. The ODPG project installed water pumping stations so that most areas had permanent irrigation. Crops became modified from cotton which was previously compulsory but of low profitability, and from traditional sorghum, to high profit vegetables including corn, beans, and eggplant.

Incomes in the project zone were found to be 137% of the control yet no positive nutritional effect of the project could be determined from the nutritional indicators used: anthropometry, haemoglobin, and food intake. This may have been due to a failure to improve intakes in line with increased income, or to a deleterious environmental effect of irrigation. Of the adult population measured, 74% had intestinal parasites in the project zone, compared with 28% in the control. Milk consumption was less in the control area as the new cropping pattern reduced forage therefore the keeping of livestock diminished. Land became expensive in the project area so that landless labourers no longer had access to buy land.

The example illustrates several issues: the question of methodology for rapid evaluation as the sample size and indicators of nutritional status used may have been inadequately sensitive to measure nutritional differences; and the importance of investigating impact beyond direct measurements on direct beneficiaries of an intervention.

Kitchen gardens

These have been encouraged in homes and schools to improve nutrition through vegetable and fruit consumption. There is little evidence that they significantly increase energy supply; however, in relation to micronutrient and in particular vitamin A status there is evidence that they can be effective. For example in Bangladesh, households with even small gardens had less vitamin A deficiency than those with no gardens (Cohen *et al.* 1985).

Common constraints include the cost of fencing, lack of water and gardening knowledge, and insufficient land and labour. Many projects have failed because they did not take into account the time availability of the intended participants. Community gardens within which individuals have private plots can be viable options and have the dvantage of facilitating provision of inputs and extension services.

Feeding programmes

Supplementary feeding programmes

Supplementary feeding programmes for young children in developing countries were reviewed by Beaton and Ghassemi (1982) from the analysis of 200 take home food distribution programmes, supervised feeding programmes, and nutrition rehabilitation centres. Such programmes are very widespread and are a component of the majority of nutrition programmes as food is usually made available through food aid and they have the advantage of being visible and directly addressing undernutrition.

However, the authors found that although they account for a substantial amount of money and food, the measured benefits in terms of growth were surprisingly small. Most claimed beneficial effects, but in some studies no effects were demonstrable. The greatest response was in those with the greatest weight deficit at entry, in children aged 4–24 months, and where there was assurance of continuity of food supply. They concluded that such programmes are expensive for the measured benefits, but cautioned that the measurements made are usually restricted to anthropometry which may not reflect the total benefit to the child or family.

Other benefits have been measured in a few studies. Morbidity and mortality have been shown to improve, while increased activity may result in improved cognitive development before measurable benefits in growth can be detected. No evidence has been found of effects on breast feeding. Leakage of food from the intended beneficiary has not been traced but this is not necessarily detrimental as it could increase the effective income of the family.

Some reasons for the small beneficial effect include the age of the children, the possibility of cross infection in on-site feeding programmes, the cost in maternal time, non-compliance with set eligibility criteria so that programmes include children that are not malnourished, and the problem of acceptibility of certain donated food products.

The authors conclude that although widespread, the present scale of young child feeding programmes is too small to have a major impact on communities or countries. There is a need to define the true objectives better, whether the intention is to improve growth or cognitive development of the child at risk or to act as an income transfer to the family. The timing of supplementation is important to consider as responsiveness to interventions may depend on age, so that if enhanced growth is the main objective children should be targeted during the first year of life.

School feeding

Malnutrition can affect school performance by reducing enrolment, attendance, attention, and achievement. In the health sector the needs of school children cannot compete with the more vulnerable under-fives and so they are often neglected. School feeding programmes are a means of targeting this age group. They are unlikely to have an effect on growth but may have on educational performance (Levinger 1989).

School feeding programmes need to be self-sustaining and relevant to the actual nutritional problems. A contrasting example was that of Iran where in 1974 a royal decree established a school feeding project, without initial consultation with the Nutrition Institute. This consisted of 500 ml milk and a cake at midday for all children in the first nine grades. Later changes occurred in the programme as the supply of milk was inadequate, and where it was available, children were not used to drinking it. The allotted quantity was therefore halved to 250 ml and later regional decisions were taken on the daily ration of 300 kcal. A typical week's menu for each day was:

1) almonds; 2) cake; 3) cheese and biscuit; 4) biscuit; 5) pistachios; and 6) milk powder.

This food supplied energy but did not address the specific deficiencies that were known to be widespread: vitamin A, riboflavin, iodine, and zinc. It was not targeted to the needy, therefore very expensive, costing US $200 million per year and more effective use could have been made of this money. However, the choice of intervention in this case was not based on assessment of actual nutritional need, but was an action designed mainly for political visibility.

Food for work and employment provision schemes

General employment policies can eventually affect nutrition but direct labour-based interventions such as food for work and employment provision schemes can be targeted at vulnerable groups to improve household purchasing power and hence nutrition, although such programmes rarely have specific nutritional objectives. The stated purpose is usually to provide employment, income to the poor, or infrastructure. They can have the long term benefits of producing assets for the community through public works such as roads, irrigation channels, drainage, and buildings. They may be seasonal, and thus effective in smoothing seasonal fluctuations in income and consumption, and preventing the sale of assets during seasonal hard times. Wages are given in cash or in kind, often using food aid. The food provided represents an alternative resource transfer, rather than a specific nutritional feeding programme. As the wages provided are usually below the market rates, the programmes are self-targeting to the poor. Projects are normally at a specific location and therefore often require travel so that the net income of the beneficiary is the difference between wages and the costs of travel, accommodation away from home and the loss of any income from production at home. However, diversion from home production is unlikely to be a problem as most participants are landless.

Food price policies

The price of food has a profound effect on nutrition as it determines the quantity and pattern of food purchases of consumers, and the income and hence consumption of producers. A policy dilemma therefore exists between the production incentive of high food prices, and the consumption incentive of low prices. Several measures have been used to subsidise food consumption as a form of social safety net in the absence of a developed social security system. Those described here, food subsidies, food rations and food stamps, are all means of lowering the real price of food for groups of consumers.

General food subsidies

In this case the government pays part of the production, storage and marketing costs of food, generally grain staples. No explicit targeting to poor or malnourished sections of the population is involved although in practice the existing marketing channels affect distribution. The influence on nutrition can be effected by: 1) increase in food purchasing power; 2) encouragement of households to buy more food; and 3) changes in diet composition. Such programmes have tended to improve the real incomes and nutrition of the urban rather than the rural poor, and this is often the primary objective, to reduce the urban cost of living and prevent urban unrest. However, they are expensive: in 1975 they took up a large proportion of government budget expenditure in many countries: Egypt 21%; Korea 19%; Sri Lanka 16%; and Morocco 12% (Berg 1981).

When such programmes are cut back the effects can be detrimental both politically and nutritionally. Food riots and increased malnutrition and death rates have occurred in several countries. However, general food subsidies are expensive and difficult to implement in rural areas, and richer households benefit more than poorer. They are therefore not considered to be a sustainable, cost-effective way of improving food security for the poor.

Targeted subsidies

Recently there has been more emphasis on targeting subsidies to the poor to cut programme costs, especially as part of measures to soften structural adjustment programmes. Subsidies can be targeted by community, household, season or commodity. The administrative burden of targeting is less on a geographical, community basis than by selection of eligible households. In addition, if households are targeted on the criteria of child anthropometry this can be a disincentive to good child care. Targeting by commodity can be done cost effectively by increasing the price of preferred grains such as rice to compensate for subsidising coarser staples such as maize and cassava, selectively consumed by the poor. Targeting by season has been relatively unexplored but could counteract seasonally variable food prices.

Rations and quotas

Rationing and quota systems can assure a more equitable distribution of a subsidized commodity. Ration schemes have been shown to be more beneficial to energy intakes than the equivalent transfer of income. In the Sri Lankan food ration shop scheme, the subsidy contributed the equivalent of 16% of the purchasing power of low income families. Outlets tend to have an urban distribution bias, as in India where only a few states such as Kerala, Jammu and Kashmir have extensive regular rural distribution; however, they can be strategically located in poor neighbourhoods.

Food stamps

Food stamps are cheaper to administer than commodity rations but as they have a currency value, they are vulnerable to inflation. Food stamp programmes have been used in several countries including the US, Sri Lanka, and Jamaica. In the latter targeting of the food stamp programme was introduced in 1984 providing a bimonthly allotment of US$7.5 worth of stamps for the purchase of cornmeal, rice and dried skimmed milk to eligible beneficiaries including pregnant and lactating women and children under five who receive health care at public health clinics, the elderly poor, very poor families, and persons enrolled in one of the other welfare programmes. The effects of targeting in comparison to a general subsidy reduced the cost from 3% to 1.6% of government expenditure, the percentage of the transfer going to the poorest quintile was increased from 14% to 31% and to the richest quintile reduced from 26% to 8% so that 51% of the poorest households were covered with only 6% of the richest households also being covered. Although food stamp programmes require good administrative capabilities, in general they are better than direct income transfers which are politically unpalatable, liable to fraud, and do not necessarily increase food expenditure.

Thailand and Philippines case study

Another example of the nutritional effects of food prices is a comparison between Thailand and the Philippines (Geissler and Miller 1982), summarised in Table 3.3. The two countries were chosen for comparison as they were very similar in regard to climate, population and economic profile, yet the prevalence of malnutrition, as indicated by the weight and height of school children, the

TABLE 3.3. Comparison of Thailand and the Philippines

	Thailand	Philippines
SIMILARITIES:		
GNP 1978 ($ per capita)	490	510
Food supply adequacy	105	108
Economic structure: % GDP from:		
Agriculture	27	27
Industry	27	27
Population (millions)	44.5	45.6
NUTRITIONAL DIFFERENCES:		
Preschool: normal weight for age (%)	44	21
School children: av. height (cm)		
boys 10 yr	130.6	125.4
18 yr	166.4	162.6
girls 10 yr	132.1	126.2
18 yr	154.6	151.2
Anaemia: whole country (%)	35	51
pregnant women	50	85
Infant mortality: World Bank	68	65
ESCAP	35.5	53.3
WHO	25.7	62
DESPITE:		
Doctors (per million population)	122	362
Nurses (per million population)	282	327
Safe water (% population)	22	39
Adult literacy (%)	64	87
Nutrition programmes ($million)	5	67
CAUSES:		
Below poverty line (% population based on minimum diet cost)	21	60

Source: Geissler and Miller (1982)

prevalence of anaemia, and infant mortality rates, were much greater in the Philippines than in Thailand. This was surprising in view of the more favourable statistics for the Philippines with respect to a variety of factors normally associated with nutrition, including expenditure on nutrition programmes, adult literacy, safe water supply, and medical services.

The main cause identified for this difference was in the cost of food. Calculations based on the minimum cost of an adequate diet showed that in the Philippines 60% of the population fell below the poverty line in comparison with 21% in Thailand. These differences were not due to different patterns of income distribution as the Gini coefficients of the two countries were very similar. They were therefore related to the price of basic foods.

Child care

Increasing household income is not enough to provide nutritional benefits (Marek 1992). In some cases maternal workload can have a deleterious effect on child care and nutrition despite the extra family income provided. In one study (Rabiee and Geissler 1990) it was found that the children of mothers with a heavy workload outside the home were more malnourished than those with a light load (Table 3.4). One of the explanatory factors for this difference was the use of sedative drugs, proprietary and opium, to keep the children asleep so that the mother could have a good night's rest (Table 3.5). Only those with a heavy work load resorted to drugs. Those children who received sedatives had a lower supplementary food intake and were both more wasted and stunted than those without sedatives.

Recent analysis has focused on the importance of women's control of resources and the caring capacity of household members who are potential caretakers of children. In practice the main responsibility for child care rests with the mother who is also often an important income earner. Attention is now being increasingly paid to means to improving her control and caring capacity.

One means is the provision of day care for children which is especially useful where labour demands are seasonal. If day care facilities are well organised they can have a direct benefit on nutrition as the children are better fed and less neglected while their parents work. Several successful examples exist. In the Punjab, harvest season

TABLE 3.4. Malnutrition factors in children < 2.5 yr related to mothers,' work load in summer (T = toman = 10 rials)

	Mothers' work load	
	Light (*n* = 18)	Heavy (*n* = 108)
Family income (T/month)	3285	3690
Diarrhoea (%)	7	16
Drug use (%)	0	30
Wasting (< 90%) (%)	17	35
Stunting (< 95%) (%)	45	39
Energy intake (% RDI)	69	65

day-care centres attached to the Narangwal Project child health programme were effective. Construction sites in several Indian cities have mobile creches managed by trained 'child care educators'.

In Colombia the Community Child Care and Nutrition Project started in 1987, sponsored by the World Bank, cares for 2–6 year old children of the poorest 20% of the population. Mothers select a 'community mother' for 15 children in her home. She receives training, a small stipend and credit to upgrade hygiene and safety. Food is provided by the government at 80% of daily nutritional requirements. Care also includes preschool learning activities. The programme is popular and had covered 500 000 children by 1989, that is half of Colombia's at risk children, at a cost of $11 per month per child.

Nutrition education

Nutrition education is a feature of most explicitly nutritional intervention programmes. In the past many nutrition education programmes have been unsuccessful as they were based on an inadequate understanding of the causes of malnutrition and were not always related to programmes that addressed economic and social constraints. They are still frequently based on out-dated information, in particular with a unique focus on protein.

However, new techniques have resulted in more effective interventions to bring about behavioural changes to improve nutritional status. The need for education is greatest where large social changes are occurring that significantly affect the amount and source of

TABLE 3.5. Malnutrition and use of sedative drugs in children < 2.5 yr during summer

	Energy intake		Nutritional status		
	Mean (kcal)	RDI (%)	Wasting < 90%wt/ht (% pop)		Stunting < 95% wt/age (% pop)
Sedative	748 + 292* (n = 22)	58	35	(n = 37)	46
No sedative	890 + 190* (n = 72)	69	30	(n = 89)	38

*$p < 0.05$ (Student t test)

income and the availability of food and non-food items, such as migration from rural to urban environments, and changes from subsistence to cash crops.

The nutrition education programmes that have produced the most positive results are those focused on changing behaviour which have been designed in association with the potential beneficiaries using both mass media and interpersonal contact rather than a simple transfer of information imposed by nutrition specialists. A common feature is growth monitoring followed by advice and discussion carefully designed for the age and status of the child.

Specific health and nutrition education in school curricula have also been shown to influence not only knowledge but also behaviour to improve health. A recent development is the concept of child to child education where children are taught health and nutrition practices to apply to their siblings. Such interventions are likely to have long term intergenerational effects.

Integrated programmes

In view of the synergistic relationship between nutrition and infection, many nutrition programmes combine interventions using health and specifically nutritional components, and sometimes include components with an indirect nutritional impact such as income generating activities. Nutrition programmes in the traditional explicit sense typically include the following components (frequency in parentheses): nutrition education (93%); health related services (73%); supplementary feeding (80%); growth monitoring (67%); micronutrient supplementation (27%); and home gardens (20%) (Gillespie and Mason 1991). Examples include the Integrated Child Development Services (ICDS) in India, the Tamil Nadu Integrated Nutrition Programme (TINP) and the Korean Applied Nutrition Programme (KANP).

Integrated Child Development Services, India

The ICDS programme, initiated in 1975, is centrally sponsored, and state administered through community centres. It consists mainly of maternal health care in pregnancy and growth monitoring and supplementary feeding for children, with the objective of improving the nutrition and health of 0–6 year olds and enhancing the capabilities of the mothers.

Health services include: 1) immunisation for all children under 6 and for pregnant women; 2) health check-ups; antenatal and post-natal for women, and for all children under 6 years of age; 3) a referral service for serious ailments, nutrition and health education to all women; and 4) non-formal preschool education for children between 3 and 6 years.

In addition, supplementary feeding is provided to selected beneficiaries. It is targeted to the most needy and malnourished, to women belonging to families of landless labourers, marginal farmers, scheduled casts and very low income groups, and to children enlisted according to their mid-upper arm circumference. Those with moderate malnutrition are given a dietary supplement of 300 kcal, with those with severe malnutrition 600 kcal per day, both for 300 days per year. Pregnant women, from the third trimester to 6 months after delivery, receive 500 kcal in the form of either ready to eat supplements or local cereals cooked on site. Additional supplements of vitamin A, iron and folate are provided to eligible participants.

In practice the targeting criteria are little used and the malnourished are reached more by judicious location of the projects. Coverage in 1985–6 was low (1.5 million mothers and 6.5 million children) and often excluded the poorest who cannot dress their children suitably to attend. Hundreds of impact evaluations of the ICDS have been carried out since its inception and the findings are often conflicting. A summary review (Subbarao 1989) indicated that the ICDS has immense potential for reducing malnutrition in India but that it has suffered from weaknesses in targeting the beneficiaries, in the recruitment and training of core workers, and in eliciting community participation.

Tamil Nadu Integrated Nutrition Programme

The TINP started in 1980, implemented by the Government of Tamil Nadu, with World Bank support. The objectives were to reduce malnutrition and the consequent high mortality of children under three years, and to improve the health and nutrition of pregnant and lactating women. By 1989 it covered 9000 villages with a total population of 10 million. A community worker was deployed in each village, supported by instructors, supervisors and nutrition officers. Operational objectives were: 1) regular growth monitoring of children; 2) short-term food supplementation; 3) strengthening of

TABLE 3.6. Economic indicators: Korea and Iran

	Korea	Iran
GNP ($ per capita): 1960	350	550
1978	1160	2160
GNP growth % 1960–78	6.9	7.9
GDP growth: 1960–70	8.5	11.3
1970–78	9.7	7.4
Distribution of GDP (%):		
Agriculture 1960–70	40	29
1970–78	24	9
Industry 1960–70	19	33
1970–78	36	54
(Manufacturing) 1960–70	12	11
1970–78	24	12
Population 1978 (millions)	36.6	35.8
Urban population (%) 1960	28	34
1980	55	50

Source: World Bank World Development Report 1980

health services as support for the nutrition effort; 4) education on child care; 5) and feeding, monitoring and evaluation.

The annual per capita cost has been estimated at $0.81 compared with $1.49 in the ICDS, the low cost being due to community participation. It has also had a greater impact on child nutritional status, severe malnutrition being halved over the four years to 1989 from 15–20% to 8–9%, with a reduction in every project area. In contrast over the same period the ICDS reduced severe malnutrition by a quarter, but not in every area. The main differences between the two programmes are the coverage of beneficiaries, the involvement of mothers, and the recruitment and supervision of workers. The World Bank concludes that free-standing nutrition projects should be narrowly focused and consist of no more than three or four well integrated nutrition interventions that do not require extensive managerial skills.

Korean Applied Nutrition Project

A comparative study between countries similar to that made between Thailand and Philippines was conducted between Korea and Iran (Geissler 1993). Over the period of comparison during the 1960s and 1970s the two countries had similarities in population size, in both experiencing an economic boom with growth in GNP and GDP of almost 10%, and a shift from an agricultural to an industrial

TABLE 3.7. Heights of school children in Korea (cm)

	7 years		11 years	
Year	Boys	Girls	Boys	Girls
1962	114.6	114.7	130.4	128.7
1967	114.8	114.5	131.8	131.2
1972	117.6	116.9	135.8	134.8
1975	119.5	117.5	137.1	137.5
NCHS Standard	124.4	123.5	146.4	148.2

Sources: Korean Ministry of Education 1962–1976; NCHS Growth Charts 1976

economy (Table 3.6). Iran was somewhat wealthier and more industrialised, yet over this period the nutritional status in Korea was shown to improve markedly with the heights of school children increasing by 7–9 cm over a period of 15 years (Table 3.7). No such change was evident in Iran.

One of the factors related to the improved nutrition in Korea was the Applied Nutrition Project. It was started in 1968 in 50 pilot villages and expanded by the end of the 1970s to almost 2000, with 8000 guidance workers in agriculture, public health and education. The main objectives were to increase the production and consumption of nutritionally valuable foods, provide education to promote those foods, organise feeding programmes for vulnerable groups and school children, train ANP personnel, and integrate with other national programmes. Other national objectives included double cropping of barley and rice and the promotion of high yielding crop varieties, community feeding and day care centres during farming seasons and diversification of the diet.

In each ANP village a nutrition station was established for demonstrations, group feeding, nutrition education and the diffusion of new recipes for substitution of the staple rice with barley. The ANP functioned at the grass roots level with emphasis on village self help, not large cash inputs, and was directed by dedicated workers of the Office of Rural Development. Village funds were raised for use on a revolving basis to finance new projects. Although the number of villages involved was small in comparison with the total of 35 000 villages, it had a multiplier effect on other villages through integration with other programmes.

The success of the ANPs in improving nutrition was evaluated in comparison with control villages. The successful dissemination of the programme caused problems for the evaluation despite the extensive

TABLE 3.8. Food and nutrient consumption in ANP and control villages per head/day

Changes in food consumption in ANP villages between 1968–76	1968	1976	Change (%)
Rice	552	374	− 32
Protein foods	45	130	+189
Green leafy and yellow vegetables	64	181	+183
Calcium foods	8	38	+375
Fats and oils	1	13	+1200
Differences in nutrient consumption between ANP and control villages, 1975			
	ANP	Control	Difference (%)
Energy (kcal)	2386	2165	+10
Protein (g)	74.5	65.6	+14
Calcium (mg)	602	473	+27
Riboflavin (mg)	1.12	0.98	+14

Source: Miller 1976

data collected by the villagers. Even so large changes in dietary patterns were shown between 1968 and 1976 in the ANP villages (Table 3.8), with a large increase in the consumption of protein and calcium rich foods, green leafy vegetables and fats and oils, and a reduction in the consumption of rice. The ANP villages showed a 15–20% higher consumption of calories, protein, calcium and riboflavin than the control villages.

CONCLUSIONS

An important function of international development and research agencies such as the World Bank, the International Food Policy Research Institute, FAO, UNICEF and WHO over the last two decades has been to draw together research on the impact of policies and programmes on the economic, health and nutritional status of beneficiaries. This has provided a more solid basis for comparison of the cost-effectiveness of interventions than has been possible to date and a means to distinguish the characteristics of success. Much remains to be done to refine evaluations and programmes but from the existing wealth of evaluation studies several features of successful large-scale nutrition interventions in relation to undernutrition have been extracted (ACC/SCN 1991). Some of the features can be summarised as follows. The objectives must be based on a careful analysis of the real problem and be achievable in a timescale set within the programme design. Community involvement in the design and implementation is essential for self-sustaining success. The overall effectiveness depends on coverage and so with targeting it is necessary to consider the trade-off between efficiency and coverage. Leadership and management is very important and effective administration must accompany the appropriate mix of components. Most successful programmes included strong elements of training and supervision. Monitoring and evaluation of the process, with flexibility to modify the programme where necessary helps ensure effective implementation. The potential for scaling up from a pilot project with selected staff to a large scale operational programme which has to use existing staff depends crucially on the attitude of the workers. These established common characteristics can be built on in the planning of future programmes to maximise their success.

Similar extensive reviews of evaluation research on policies and programmes to address the nutrition-related diseases of affluence such as obesity, heart disease, diabetes and cancer have not yet been undertaken, although a start has been made, for example by James and colleagues (1988). This is an important task for the future.

REFERENCES

ACC/SCN (1991). *Managing successful nutrition programmes.* ACC/SCN State-of-the-Art Series, Nutrition Policy Discussion Paper no 7. ACC/SCN, Geneva.

Beaton, G. H. and Ghassemi, H. (1982). Supplementary feeding programmes for young children in developing countries. *Am. J. Clin. Nutr.*, **35**(4), 864–916.

Berg, A. (1981). *Malnourished people: a policy view*. World Bank. Washington DC.

Berg, A. (1987). *Malnutrition. What can be done? Lessons from the World Bank experience*. Johns Hopkins University Press. Baltimore and London.

Brun, T. A., Geissler, C. and Kennedy, E. (1990). Evaluation of the Impact of Agricultural Projects on Food, Nutrition and Health. In *World food issues*, Vol. 2 (ed.) T. A. Brun and M. Latham. Cornell University Press.

Cohen, N., Rahman, H., Sprague, J., Jalil, M. A., Leemhuis de Regt, E. and Mitra, M. (1985). Prevalence and determinants of nutritional blindness in Bangladeshi children. *World Health Statistics Quarterly*, **38** (3), 317–30.

Dreze, J. and Sen, A. (1989). *Hunger and public action.* Clarendon Press, Oxford.

Geissler, C. (1993). Stature and other indicators of development: comparisons in Thailand and Philippines, Korea and Iran. In *Food, diet and economic change past and present* (ed. C. Geissler and D. J. Oddy) Leicester University Press.

Geissler, C. and Miller, D.S. (1982). Nutrition and GNP. A comparison of problems in Thailand and The Philippines. *Food Policy*, **7** (3): 191–206.

Gillespie, S. and Mason, J. (1991). *Nutrition-relevant actions: some experiences from the eighties and lessons for the nineties.* UN Administrative Committee on Coordination, Subcommittee on Nutrition (ACC/SCN) State-of-the-Art Series, Nutrition Policy Discussion Paper no 10. ACC/SCN, Geneva.

Gittinger, J. P., Leslie, J. and Hosington, P. (1987). *Food policy. Integrating supply, distribution, and consumption.* Johns Hopkins University Press. Baltimore and London.

Gwatkin, D. R., Wilcox, J. R. and Wray, J. D. (1980). *Can health and nutrition interventions make a difference*? Overseas Development Council Monograph 13. Washington DC.

James, W. P. T., Ferro-Luzzi, A., Isaksson, B. and Szostak, W. B. (1988). *Healthy nutrition: preventing nutrition-related diseases in Europe.* WHO Regional Publications. European Series, no 24. World Health Organisation, Geneva.

Levinger, B. (1989). *Malnutrition, school feeding and educational performance.* CARE Unit for Cooperation with UNICEF and WFP no. 196. Paris.

Levinson, F. J. (1987) Towards success in combating malnutrition: an assessment of what works. *Food and Nutrition Bulletin*, **4**(3), 23–44.

Marek, T. (1992). *Ending malnutrition: why increasing income is not enough.* WB/AFD/PHND Technical Working Paper 5. World Bank, Washington DC.

Miller, D.S. (1976). *Korean Applied Nutrition Progaramme.* Consultants for UNICEF/FAO. National Center for Health Statistics (1976). *NCHS Growth Chart HRA 76—1120*, Vol 25 (Suppl. no 3). NCHS, Rockvillo MD.

Pinstrup-Anderson, P. and Hazel, P. B. R. (1985). The impact of the Green Revolution and prospects for the future. *Food Review International*, **1**(1), 1–25.

Rabiee, F. and Geissler, C. (1990). Causes of malnutrition in young children: Gilan, Iran. *J. Trop. Ped.*, **36**, 165–70.

Subbarao, K. (1989). *Improving nutrition in India: policies and programmes and their impact.* World Bank Discussion Paper no 49. World Bank, Washington DC.

World Bank (1980). World Development Report. Washington DC.

4
The International Water Decade and Beyond: New Public Health Interventions

MORAG BELL, RICHARD FRANCEYS AND MARY LIAO

INTRODUCTION

Since the late 1970s, the provision of public health facilities for low income communities has become the subject of heightened debate. The sheer numbers involved—in many cases, some 40–50% of major city inhabitants and some two-thirds of rural populations—have provided a compelling reason for renewed efforts at social and environmental engineering (WHO 1985). By supporting the principles of slum and shanty upgrading and minimum services in rural areas, the international aid community has given political respectability to various low cost approaches and technologies (Glaser 1985). This paper explores recent attempts to improve the effectiveness of environmental health programmes and projects as reflected in the International Drinking Water Supply and Sanitation Decade, or Water Decade, and beyond. Launched by the United Nations General Assembly on 10 November 1980 following the Mar del Plata Conference in 1977, this global movement sought to improve human access to clean water, appropriate sanitation and primary health care by 1990.

The chapter considers the origins of this movement as a critique of the high technology principles underpinning modern industrial society, advocates of the movement being committed to basic needs as human rights. It reviews the international political–economic context for Water Decade practices. Reference is made to the images and ideologies relating to individuals and communities as producers and consumers of services, which underpin these practices. Particular attention is given to how water and sanitation as technical interventions have drawn upon the natural sciences, notably concepts of race and sex, and the social sciences including culture and gender, for their authority and legitimacy. This paper demonstrates how, during the course of the Decade, sections of the aid community redefined basic needs as commodities involving the 'efficient' marketing and delivery of a product with minimal state intervention. Within this context, it examines the changing role of the expert and the emerging relationship between large aid donors, non-governmental organisations[1] and the private sector. Finally reference is made to the significance for the development debate of this reformulated progressivism, defined as the commodification of human rights, including the challenge it presents to concepts of the 'other', notably poor people and women, to technocentrism and to the role of the state.

TECHNOLOGY AND A HEALTHY SOCIETY

The benefits of improved domestic water supply and sanitation facilities in enhancing community health and labour productivity have underpinned infrastructure improvements in Britain and the colonies since the mid-nineteenth century. As part of the public health movement, these technical interventions have not merely sought to fulfil the practical demands of capital for an efficient labour force or the needs of the state for social control. They have been implicated in broader, idealist goals such as the creation of a 'healthy' city and a 'progressive' society (Driver 1988). Thus, in the opening years of this century, public health in Britain became an integral part of the garden city movement (Buder 1992). Together with urban planning and the eugenics movement it was incorporated into scientific management and facilitated attempts to produce a vigorous, morally-upright imperial race (Taylor 1911). In the aftermath of the South African War, 1899–1902, environmental improvements were deemed to be essential to the regeneration of British society. For social reformers and politicians these assumptions about the relations between the physical environment and citizenship were substantiated by an appeal to the apparently benign influence of science and

[1] NGOs are non-profit-making agencies either international in origin or indigenous to the developing countries, engaged in the provision of services for relief and development purposes.

technology (Rimmer 1981; Ward 1989). Physical infrastructure, based on the knowledge and skills of scientists and engineers, conformed to the image of rationality and efficiency incorporated in the machine (Harvey 1989). It reflected a new and widespread faith in enlightened principles of modern industrial society. Nor were the practical and idealist objectives underpinning social and environmental transformation confined to Britain. They became an integral part of imperial city planning and attempts at 'civic renaissance' in mainland Europe, North America and Britain's colonial territories overseas (Dossal 1989).

In the post-colonial world, the creation of an ordered landscape, efficient economies and disciplined societies, remain goals of international aid donors based in Europe and North America. Thus a central theme in the discourse of international development is the power of western management, science and technology. Large scale, costly and ambitious projects exported overseas by the aid community in the name of 'progress' are imbued by a spirit of professional and scientific superiority little different from their colonial predecessors. In the case of a technical sector like water supply and sanitation, dominated by sophisticated engineering, these 'development' assumptions are reinforced. The Mahaweli irrigation scheme, begun in the dry zone of Sri Lanka in the mid-1970s, is one example. Funded by overseas aid and involving large scale population relocation from the country's wet zone, it has involved not only costly technical investment but also the intensive re-training of farmers to alter their long-standing land and water use practices (Madeley 1983).

In the case of groups outside the formal housing sector and state control it has proved extraordinarily difficult for public or private authorities to provide sound and lasting technology and observable improvements in health. The management of low income communities has been a slow and haphazard process. In Europe and North America, water supply and waste disposal, initially installed in wealthier neighbourhoods, gradually spread to poorer ones (Headrick 1988). Not so in the colonial and post-colonial territories; rural and urban infrastructure has been largely neglected in relation to the numbers involved (WHO 1985). Women continue to walk long distances for water in rural areas, with negative effects upon their health (Mehretu and Mutambirwa 1992). In the major non-western cities, despite the use of imported technologies in municipal planning and public works, poor people continue to concentrate on land which is liable to flooding, on buffer zones close to railway lines or on steep gradients neglected by the construction industry. Confined to environments which confound the ideal development image, they remain largely marginal to the project of modernism.

Official responses to these 'other' groups have, however, been complex and varied. They reflect not only a concern with revenues and the funding of public works but also faith in a number of often conflicting assumptions about the nature of these social groups and their economic value. Underpinning these assumptions has been the concept of difference; the application of certain physical, moral and mental attributes to those who, by virtue of their race, class or gender, fail to conform to a eurocentric and intensely masculine concept, the rational, ordered world (Vaughan 1991). Deep-rooted anxieties over the deviant behaviour of these non-conforming groups or of the public health risk they posed to urban middle classes have, at various times, promoted explicit forms of segregation and control. In Singapore at the turn of this century, Yeoh (1992) notes that for municipal health officers imbued with an unquestioned confidence in the supremacy of western sanitary science, the prevalence of 'filth' diseases like cholera among the Asiatic population, particulary the Chinese, was directly related to their 'careless' and 'revolting' domestic practices. Ill health, crime, intemperance and ignorance appeared to emanate from the environments within which 'other' races were located. With the advent of industrialisation and urbanisation in South Africa, Maynard Swanson (1977) shows how from the 1870s, fears of epidemic diseases both aroused and rationalised efforts to segregate Indians and Africans in municipal locations and to define South Africa's 'racial ecology' (Marks and Andersson 1988).

In contrast to this colonial neurosis, sections of the urban poor, like their rural counterparts, have not always confounded the 'model citizen' image. Shifting boundaries of difference can be defined in the many hopeful speculations by social scientists, public, voluntary and private bodies, that marginal groups could not only be effectively managed but their value as human capital realised. At various times in the colonial and post-colonial eras, urban upgrading including the provision of public housing and basic services has reflected official enthusiasm to stabilise peripheral sections of urban society; for reasons of political opportunism, as a means of harnessing their labour power in the drive to achieve economic growth in addition to being a response to the demands of the residents themselves (MacLeod and Lewis 1988). Thus embedded in public health discourse, pessimism has complemented periods of optimism that through appropriate environmental interventions, those on the margins of society could be incorporated into the technologically-sophisticated modern world.

THE WATER DECADE AS A PROGRESSIVE MOVEMENT

The Water Decade appeared to continue this optimistic tradition; but it also sought to challenge the powerful and enduring high technology image of progress on which western environmental and social transformations have been based. In doing so, it brought together a range of interests. Encouraged by liberal thinkers and promoted by experts in academic institutions in the western world, the Water Decade movement was spearheaded by international agencies. Global support for a specifically sectoral strategy was designed to ensure the more efficient and effective management of development programmes by both national governments and the aid community. Although in many countries only a small proportion of public expenditure and aid budgets is assigned to human development (UNDP 1991), this vertical intervention provided a common focus within which governments could set targets. It attracted international donors to provide resources, know-how and better coordination of their aid programmes. In effect, it sought to assist these major institutions to improve their accountability.

But efficiency for donors and governments did not appear to be incompatible with equity and social justice. Implicit in the Water Decade philosophy was the basic needs approach to development in which access to safe water, appropriate sanitation, food and housing were regarded as essential human rights. Poverty linked to disease was once again on the official agenda. Although throughout the world domestic water is used primarily for drinking, cooking, washing and bathing, coverage is highly unequal. By improving the quantity and quality available, a positive impact on the health of children under five years was anticipated since they suffer most acutely from inadequate water and sanitation. Thus the allocation of scarce resources to this technical intervention was given authority and legitimacy through an appeal to medical science. Moreover, standards of need were no longer to be defined merely in western, elite terms or on the basis of ability to pay. Rather, the urban and rural poor were targeted as deserving cases for infrastructural improvements. It was argued that

> A concerted effort should be made to provide reasonable access to safe water supply to all communities, including remote villages, on the basis of concrete targets and timetables to be set up in each country (United Nations 1977 pp. 24–5).

The health benefits, notably substantial reductions in infant and child diarrhoea, appeared to be undisputed and entirely justified a financial subsidy.

In highlighting the material necessities of life as goals of development, the Decade formed part of a widespread contemporary critique of modernisation; it offered a challenge to the culture which, for much of this century, has underpinned western notions of progress. In seeking to redefine development as a broad-based, people-oriented process (*Decade Dossier,* undated), it captured an international concern for social amelioration and environmental quality uniting the western and non-western realms. In consequence, inner city 'problem spaces' in Britain and North America shared some common ground with the slums and shanties of Bombay and Sao Paolo (Peterson 1991). Nor did the Water Decade philosophy operate in isolation. It formed part of a broader contemporary shift in medical thinking declared at the Alma Ata conference of 1978 and enshrined in the WHO strategy of 'Health for all by the year 2000', in which primary health care, of which water and sanitation form a part, was to be given prominence over curative approaches. In effect, new opportunities were being seized across a range of complementary sectors to provide better health at lower cost (Lowes 1982).

Celebrating alternative technologies and new visions of the 'other'

In its contemporary form, the Water Decade movement was driven by a recognition that sophisticated technology could not satisfy health needs. It built upon the appropriate technology tradition of

E. F. Schumacher (1974) around which principles of pre- and post-modern design have converged. The effects of 'alternative' technologies upon urban form are undoubtedly different from the transformations in architectural design and industrial processing encouraged by new building materials, computer modelling and the communications revolution of late capitalism. Nor can they be compared with the 'nightmarish post-modern landscape of homelessness' which is the lot of the poor in the core countries (Harvey 1989 p. 77). Rather, in the slums and shanties of major cities and in rural areas, the visual impact is one of do-it-yourself diversity; an expression of the profound effects of 'alternative' technologies upon the production of social infrastructure, the delivery system, the market for social services and systems of financing.

Low-cost solutions to water supply and sanitation have revolutionised the production of these services. Communal handpumps and pit latrines offer effective substitutes for complex piped systems and avoid the reliance upon scarce, expensive imports. These alternative technologies have challenged established legislation, often colonial in origin, over what are deemed to be the minimum acceptable standards in construction and services. Changes in the sphere of production have also affected market demand. Alternative technologies cater for a wider population than previously. A more extensive, varied and fragmented market for social services could be exploited than formerly. Thus, during the course of the Decade, it was argued that all programmes must be 'innovative', 'imaginative', and 'enlightened' to reflect the variations in cultural, environmental and economic conditions within and between countries (Kalbermatten 1983).

Alternative technologies have also affected the delivery system: a shift from complex structures enables public and private sector monopolies to be broken. Responsibility for infrastructure provision can be shared with the beneficiaries themselves. Participatory approaches in turn, have implications for forms of knowledge and institutional structures. They challenge 'top-down', managerial approaches controlled by the 'expert' nurtured in western ideas of the modernist scientific enlightenment. Planning can be transformed from centrist design and mere 'hardware' projects; decentralisation of power and decision-making can be encouraged drawing upon local skills and knowledge. In the case of water and sanitation, a major challenge was therefore presented to technical assistance within international aid and to technical ministries within governments. John Kalbermatten reported in 1983 that 'some of the most rock-bottom conservatives involved in the Water Decade—engineers, economists and doctors—are actively participating in the revolution' (p. 1). Indeed Water Decade rhetoric came to bear many hallmarks of a post-modern age. It celebrated diversity rather than homogeneity; flexibility rather than bureaucratic rigidity.

These revolutionary changes also provided the opportunity to change systems of financing. On the one hand a simplified production process reduces production costs. On the other, implementation costs can be lowered by community participation. It is within this context that donors and governments have rediscovered 'marginal' groups including poor people and particularly women, as the primary users of water systems. The Water Decade derived much of its authority, legitimacy and unity from a new vision of the 'other' both at home and overseas. Representing a legitimate new market for alternative forms of infrastructure, these target groups challenged much development orthodoxy. While improved environmental health was regarded as a physiological need for human survival, the interventions advocated were built upon a set of revised assumptions about the social, moral and mental characteristics of the beneficiaries. Gone was the imagery of disease as a societal metaphor and a political weapon to justify social and racial segregation and exclusion. Also destroyed were images of the backward, lazy and idle 'poor'; mere objects of reform. Rather, men, women and children of the periphery were subjects of change, that is, active agents in their own development (Bell 1991).

In a drive to formulate and implement more people- and gender-sensitive policies, throughout the 1980s the donor community was remarkably consistent in exhorting national governments to promote community participation and the involvement of women (Elmendorf 1990). This was particularly so in rural areas where conventional engineering provided by the state was impracticable. Official 'images' of these groups were positive and progressive. From the beginning 'recipient' communities were to be encouraged to participate. The Mar del Plata Conference which initially formulated the Water Decade philosophy in 1977, recommended that countries should 'adopt policies for the mobilisation of users and local labour in the construction, operation and maintenance of projects for the supply of drinking water and the disposal of waste water' (United Nations

1977 p. 25). There was to be mass mobilisation of local institutions including community organisations, self-help groups, health clinics and schools. That this community participation should specifically involve women, the drawers of water and guardians of family health, was explicitly stated in 1980 at the World Conference of the United Nations Decade for Women. By 1982 an Inter-Agency Task Force on Women and the Water Decade had been established (INSTRAW undated). A report of 1988 reiterated the by-then-familiar platitudes:

> to be successful, rural projects should be community based, should be sustainable and replicable, and should stimulate the community's long-term commitment and support. Community means the full community—not just the leaders—and especially those who stand to benefit most and whose daily support is needed for sustainability. In water and sanitation this particularly means the women of the community' (UNDP/World Bank 1988 p. 7).

Thus the technical sector and neglected social groups not targeted. Implicit in the Decade philosophy was the assumption that by improving the physical environment of these groups, and by implication its aesthetic quality, technology which was culturally and economically sensitive, could help to reduce their marginality and bring them within the orbit of mainstream development. That these benefits could be achieved through collaborative action appcared to be entirely natural since rural communities in particular, were 'socially cohesive, committed to the common good and had valuable local knowledge which could be harnessed in project design and implementation'. The consequences for women would, it was argued, be particularly great. By reducing their time involved in collecting water, new opportunities would be available to engage in more productive tasks and to contribute directly to economic growth. Equally, since appropriate technology would help to reduce infant mortality rates and, by implication, the need for insurance births, fertility might also decline. Thus the watchword of this new environmentalism was improvement and reform through appropriate social action from within. The benefits appeared to be infinite; equity and social justice, environmental quality, economic growth, improvements in population quality and perhaps indirectly, reductions in population growth.

Much of the discussion which follows addresses the water sector only.

THE BROKEN HANDPUMP AND THE EMPTY TAP AS METAPHORS FOR DEVELOPMENT

The outcomes of the social and technological 'revolution' described in the previous section are variable. Many quantitative evaluations were carried out during the Water Decade which indicated some improvements in coverage (WHO 1985). By 1990, 82 per cent of urban dwellers and 63 per cent of rural people in the lower-income countries were believed to have a suitable water supply (UN 1990). However, there were wide variations within and between countries. More difficult to quantify, and more disturbing for funding agencies, were the many facilities unused, inoperable or poorly maintained. Also of concern were the uncertain health impacts (Esray *et al.* 1985). Communities had proved to be unexpectedly slow to accept, to become involved in, or to benefit from, apparently enlightened strategies.

Explanations vary reflecting the interests and institutional loyalties of the participants. Within the aid community reference has been made to rapid population increase within countries and to the lack of commitment by recipient governments. External funding, they continue to argue, must depend upon comparable national support for the sector (Dieterich 1983). Limited human resources and management skills within governments have been recurring themes; hence the need for institution building (Beyer 1988; Brown 1985; Franceys 1991). Recipient governments have stressed limited funds. By 1983, 'Whatever happened to the World Water Decade?' was headline news in many countries, spreading alarm that it had been abandoned by donors (Dieterich 1983). At the same time, poor operation and maintenance has frequently been blamed on the uncoordinated proliferation of donors and hence of technologies (WHO 1985: 49; WHO 1988: 5). The technology itself has also come under attack promoting research into new designs (Arlosoroff *et al.* 1987).

However, despite these differences in emphasis, development planners, sectoral specialists and project practitioners have been tempted to apportion some of the blame, either implicitly or explicitly, to the beneficiaries themselves (Huston 1985). In most cases they have simply failed to conform to the cultural model of progress on which appropriate technology was based. In rural areas organisations like UNICEF, Scandinavian aid and NGOs have made enormous efforts to promote participation. Nevertheless, lack of sustained local commitment, indifference and ignorance have, many claim, contributed to the failure to meet Decade targets. 'I tried to get women's participation, but they wouldn't come to my meetings'; 'We've "done" women in our agency—there is a policy statement which says women are important and we have a WID (Women in Development) office. Now we want to get on to other matters'; 'How do we "do" community participation?'. These are some of the attitudes and observations voiced (Melchior 1989). Equally, among the 'beneficiaries' health benefits appear to have been far less important in promoting participation than privacy, particularly for women, convenience and being 'modern' (van Wijk-Sijbesma 1985). In effect, there has been a reluctance to accept the environmental and social 'improvements' on offer in the terms defined from outside.

In consequence, investment strategies by major bilateral donors and development banks have maintained the historic bias in service provision towards urban middle and upper class consumers (Biswas 1987). In 1991 it was reported that some 70–80 per cent of funds were used to serve a mere 20–30 per cent of the population (Kalbermatten 1991). These were mostly the wealthy with access to formal housing and the political power to secure from development banks, via recipient governments, sufficient capital expenditure for conventional technology; household water connections and sewerage continue to be provided, often at hugely subsidised rates (Christmas and de Rooy 1990).

As the international policy environment changed during the 1980s, so too did the discourse. The Water Decade opened with an emphasis on concrete goals and timetables. By mid-term, optimistic national and global targets had been revised downwards (WHO 1985). Initial statistical objectives were being redefined to focus on the less tangible aims of raising local, national and international awareness (*The Courier* 1986). In the words of George Arthur Brown, UNDP Associate Administrator and Chair of the Steering Committee for Cooperative Decade Action, 'The most important effect of the UN's involvement in the Water Decade, is that we have succeeded in great part in building up an awareness by governments not only that the Decade is important, but that it is possible to do something about it' (*Decade Watch* 1984). Furthermore, by mid-Decade, the provision of water and sanitation for low-income groups was no longer sector specific. It had become increasingly integrated with other sectors either as part of a package with housing, or within integrated programmes like rural development or primary health care (*Decade Watch* 1984). Indeed, for health workers, water supply was proving to be an excellent 'entry point' to communities; evidence which encouraged many donors and national governments to encourage the 'softer' ministries with community level experience—Health, Community Development and Women's Affairs—to champion low cost water supply and sanitation in preference to the apparently more inflexible technical staff within the 'hardware' ministries.

Pluralism and the commodification of human rights

Recent documents reviewing the latter part of the Decade and beyond, reflect two trends in particular, namely, a drive for economic efficiency and, coupled with this, the need for alternative methods of welfare financing and provision. As donors confront the rigours of accountability at home, so pressure upon them has intensified to improve operation and maintenance where the record has been poor. To achieve this, service provision has become increasingly linked with the culture of the market. During the course of the Decade, the UNDP/World Bank, taking its lead from the 'new' political Right, led the way in ensuring that the concept of basic needs, codified internationally as legal rights, was narrowed down and its meaning altered by a succession of additional labels including willingness to pay, cost recovery and hence replicability (Churchill 1987; Briscoe and de Ferranti 1988). In effect, basic services as an essential part of collective consumption were transformed from human rights to commodities subject to the rigours of the market. Thus the goals at the beginning and end of the Water Decade looked somewhat different. 'Need' became shaped by cost recovery. The new language was 'effective utilisation' and 'sustainability' (Churchill 1987; Narayan-Parker 1989*a*).

This donor emphasis on economic efficiency has assisted recipient governments which confront national debt, structural adjustment with its associated cuts in public expenditure, and bureaucratic mismanagement (Stewart 1991). In circumstances where the public sector cannot provide, alternative methods of welfare provision involving new approaches and strategies become intensely attractive. They include three in particular: increasing emphasis on community management and with this, a redefinition in the roles of the expert; greater pluralism in the organisations involved within public and private sectors; and renewed commitment to better coordination of the aid effort. These approaches are closely linked and are based on a new set of assumptions, both practical and ideological, about the roles of communities, governments and donors.

Community management

Implicit in Community manegement is the assumption that community participation—often a convenient euphemism for unpaid labour—is not enough. Community management is based on a consensus model of society which is simplistic and naive, namely, that external agency and community values and priorities coincide and that the latter, given the necessary motivation through education and skills training, will voluntarily participate in externally-induced environmental and social engineering. In effect, there is a single model of development to which everyone aspires. While not wishing to advocate a crude conflict model of society—that broken handpumps are an expression of local resistance to external control and coercion—what donors and governments have frequently failed to appreciate is that the communities they seek to mobilise remain marginal to much of the decision making which affects their lives. During the Decade, important research and advocacy were carried out by the United Nations Development Programme (UNDP) /World Bank Water and Sanitation Program and by PROWWESS (Promotion of Women in Water and Environmental Sanitation Services) (Report 1991). But despite rhetorical support for a 'software component' in many projects, it was frequently neglected, poorly implemented or largely ignored in project design (Yacoob 1988). In effect, projects advocating institutionalised participation remained of the familiar top-down, externally-controlled type (Wisner 1988).

Community management principles adopt a position between the consensus and conflict models. The concept implies that community members are active decision makers taking responsibility for, and control over, decisions affecting how services are provided and maintained (Arlosoroff *et al.* 1987; UNDP/World Bank 1989). In shifting the emphasis from technology and sustained functioning of services to their effective and sustained utilisation, it is argued that devolution of power, authority and control is essential (McCommon *et al.* 1990). The concept suggests much more than self-help labour. It implies that communities are well able to organise and manage change. They are not deviant, apathetic or ignorant; rather, as rational entrepreneurs, managers and consumers, they are well able to take over a measure of responsibility for service delivery and maintenance. The social action advocated builds on the principles of active citizenship, namely, individual duty and collective responsibility.

The role of the professional

What then is the role of the professional? Community management challenges much of what is implicit in the work of the technical expert, namely, the uniformity and inflexibility of sector-specific and time-bound projects. It involves providing the conditions on which the effective and sustained utilisation of services depend. These include promoting modern leadership, organisational skills and local decision-making structures—human resources development—so that communities can take an active part in decisions about the services provided (Narayan-Parker 1988; 1989*b*). In effect, it involves strengthening or reshaping the social organisation of 'marginal groups' in order to achieve sustainable and replicable projects. The locus of action shifts from 'initiation and mobilisation' of communities to 'responsibility and participation' (Donnelly-Roark 1987). Achieving this requires social skills in order to determine the workings of communities and their own priorities (Narayan-Parker 1989*a*). It also requires time; interpreted here as a resource not a constraint. As Quratul Ain, community organiser in the sanitation upgrading of Baldia township, Karachi, points out, the first four years of the project (1979 to 1983) served as a learning base in which social distance between community and outsider was gradually narrowed (Bakhteari 1987).

The role of voluntary organisations

Ensuring that community management takes place is acknowledged to be a slow and complicated process. Lacking established models and rules, it is not amenable to standardised technical and managerial techniques. Highly labour intensive and costly for officialdom in terms of the time involved, it conflicts with technocratic efficiency such as the drive to meet construction schedules or to use up aid funds. It is therefore very difficult for bureaucratic organisations to accommodate. In these circumstances, it is argued that governments and donors should no longer be seen as the monopoly providers of services; there is a role for many 'other' organisations within the private and voluntary sectors. Non-governmental organisations are highlighted as particularly well placed to promote community management. Their approach to projects is less formal and more flexible. They are not bound by the rigid project cycle. They have the capacity and commitment to encourage marginal groups including women and the poor and to act as intermediaries between governments and the people; assumptions which are not always realised in practice (Yudelman 1987; Yacoob 1984).

Given the variety of local social, economic and environmental conditions, there can be no single model of service delivery. Rather, the form and scale of participation by 'other' organisations in relation to government control will vary according to local circumstances. But the national context is also important. It is for the central state to provide the 'enabling' environment within which alternative models can flourish. Thus diversity and innovation are envisaged, demanding of the expert a willingness and flexibility to work with a plethora of service delivery models within a supportive political environment (World Bank 1989). Ironically there is nothing new about the pluralism advocated. In most developing countries the role of the state as provider of collective consumption has always been extremely limited. The vast majority of people have long had to find their own solutions to housing and basic services.

Coordination and international commitment

But pluralism, according to the major donors, also requires cooperation (Briscoe and de Ferranti 1988). At the International Drinking Water Supply and Sanitation Consultation held in Interlaken, Switzerland, in October 1987, it was agreed that 'sustained' development of the sector would benefit from more 'efficient' and 'rational' cooperation between aid agencies, including NGOs, during the remainder of the Decade and beyond (UNDP/World Bank 1988). In support of this rhetoric, major donors including UNDP, UNICEF, WHO and World Bank, have enthusiastically sought to collaborate with NGOs (Qureshi 1988; Timpson 1988, 1989; UNFPA 1988).

Collaboration is not, of course, new. What is new is the international donor commitment to it and the policy environment which has stimulated it. Nor is it unique to the water sector. Since the early 1980s major multilateral and bilateral agencies have become champions of NGO activities in a range of sectors including agriculture and industry (Drabek 1987; Fowler 1992; World Bank 1988).

A similar commitment to coordination was expressed in September 1990 at the Global Consultation. Staged in New Delhi to debate the end of the Water Decade and to define goals and procedures for the 1990s, this cosmopolitan gathering of advisers, field workers, pressure groups and critical observers was seen as a 'unique historic event' (Rothermel 1990). Present were women and men of many realms—multilateral and bilateral donors, technical, planning and social development ministries within recipient governments, non-governmental organisations, development banks, consulting engineers and academics. In an environment of economic stringency, the staging of such a grand and costly gathering which looked forward to the 1990s, was important in itself. It sought to ensure that social infrastructure remains on the political agenda. It also symbolised the donor community's new humility; a willingness to acknowledge institutions other than formal bureaucracies; forms of knowledge other than that of the expert, and diverse models of service delivery. One tangible outcome is the establishment of a Water and Sanitation Collaborative Council to foster, at regional and global levels, the cooperation developed during the Water Decade (Grover and Howarth 1991). Under the aegis of the Council, an Information, Education and Communication section has been set up to provide an integrated communication strategy promoting a range of activities for policy makers, sector staff and users.

TECHNICAL INTERVENTIONS AND THE DEVELOPMENT DEBATE

The changes discussed in the previous section, raise many issues which are central to development theory and practice. They challenge the concept of development as progressive evolutionary change. Official dialogue with voluntary agencies indicates that while the welfare mix overseas has changed during this century to one in which the public sector has increasingly taken over responsibility from family, community and charity, this is not an irreversible process. There is no smooth and linear expansion of the sector over time from voluntary to public control. Nor is there a smooth process by which marginal groups are materially incorporated into the dominant culture. The process is more complex and varied. Active support for NGOs spreads responsibility at a time when public resources are scarce. A changing role for government is clearly envisaged 'from that of provider to . . . promoter and facilitator' thus enabling 'local public, private and community institutions to deliver better services' (New Delhi Statement 1991 p.116).

As responsibility shifts away from planners and bureaucrats, there is a need to re-evaluate the assumptions on which these technical interventions are based. To what extent are equity and efficiency compatible objectives? What is meant by popular participation? How does a supportive political environment operate in practice? Does inter-agency coordination bring a return to centralism? Crucially the evidence suggests that we must move away from monolithic and homogeneous categories.

On active citizenship

In adopting a non-interventionist approach, donors are admitting in both political and managerial terms, that there are certain localities where the state can play only a minimal role. This retreat from service provision, except for wealthy urban dwellers, would appear to marginalise low income groups by deliberately excluding them from the benefits of state services—echoes of the undeserving poor. Admittedly government inertia is officially condemned; the state, it is argued, has a vital role to play in providing 'the enabling environment of appropriate policies, legislation, incentives' and 'extension' (New Delhi Statement 1991 p.116). But this is a suitably vague exhortation and overlooks individual expectations of government as the provider of basic human rights. It raises important questions about the rights and duties of citizens, the organisation and control of public services and the broader management of urban and rural affairs. Why should poor households in one region of a town or country be required to purchase and manage their own services while the state provides for the wealthy? Undue emphasis upon the duties of citizens diminishes the importance of enabling them to define and assert their rights. Evidence suggests that for low income groups, the costs in terms of time, energy and money of purchasing and managing services can be very high compared to those who pay the state as provider and manager, often at heavily subsidised rates (Briscoe 1992; Cairncross and Kinnear 1992).

In the case of many African countries, a contradiction may be emerging in contemporary aid policy. At a time when donor agencies are promoting citizen duty and responsibility in service provision, the political system is becoming more open as large donors insist upon democracy as a precondition for external support. New opportunities may therefore be emerging for low income groups to exert pressure on the state by articulating a range of citizenship rights, including their rights to services.

On local democracy and local knowledge

In promoting communal action a sense of duty and individual responsibility is called upon based not on secular legislation or on local state monopoly but on principles of reciprocity and social cohesion supposedly persisting among poor communities in the modern world. 'Measures to strengthen local institutions' are nevertheless deemed to be essential (UNDP 1990 p.2) to encourage democratic localised control. But which institutions? While to the 'outsider', communities may display a mask of simplicity, internally they are complex and dynamic, neither homogeneous nor harmonious, the actions of whose members may be driven by a variety of motives other than collective responsibility (Jones and Wiggle 1987). Many barriers exist to a democratic and participatory form of citizenship. The scale at which collaborative action works effectively may be crucial. Experience of the Decade indicates that with the exception of small projects based upon personal loyalty, civic duty and a cooperative spirit, community participation is often difficult, time-consuming and does not necessarily save money. In rural Indonesia the arisan system, a traditional community saving system involving a small group of friends, has been effectively adapted to water and sanitation projects. In Galle, Sri Lanka, a cluster of 20–40 households has proved to be an effective grouping within which responsibility can be shared on a cooperative basis for managing a distribution network and collecting revenues (Cotton and Franceys 1991).

And what of community mobilisation and the sustainability of community-managed projects? Individuals are being exhorted to act both as rational consumers and to take communal responsibility for the maintenance of services. They are, in effect, being asked to embrace what may prove in practice to be two mutually incompatible principles, namely, individual liberty and collective responsibility. To what extent can individuals be encouraged to pursue their own private goals and be exhorted to act collectively for the common good? Evidence from Zimbabwe indicates that effective mobilisation frequently depends upon the initiative and energy of a few key individuals (Boydell 1990). Difficulties arise in sustaining this initiative when they move on. While groups of people may join together for a specific short-term goal like laying a trench or building a latrine, thereafter they want to get on with their lives. Long- term sustainability appears to depend upon establishing formal arrangements with government and/or NGOs (Trace and Whiteside 1992; Wood 1992). In adjacent South Africa where pressure for improved social services is now great, how to promote social citizenship among the majority of the population as a means of meeting increased demand, is a pressing contemporary question. It is especially so in circumstances where that population has hitherto been denied political representation and the right to control their lives and where expectations of the post-apartheid state are high (Daniels 1992). Equally, strategies to decentralise control should not obscure the fact that communities are arenas for conflict, negotiation and compromise. As the custodians and mediators of local knowledge, political elites may retard the market or discourage collective action in particular settings. Davies (1990) demonstrates with reference to rural water supplies in Lesotho that local politics, notably conflicts between different groups at village level, together with a lack of specific items of knowledge by members of the village water committees, constrained the effective management of the supplies.

The paradox of participation

Greater realism in interpreting the dynamics of communities needs to be matched by a close look at household structures and gender roles and relations within different societies. Recent gender development critiques (Moser 1989; Rathberger 1990; Sen and Grown 1987) provide a framework for analysing existing initiatives in the water sector (Liao and Rogers 1993). In terms of policy, women's role in improving the quality of collective consumption remains unchallenged; their full participation continues to be encouraged 'at all levels in sector institutions' (UNDP 1990 p.2). But there remain few women in positions of professional influence. In consequence, men continue to predominate in national and international fora discussing issues like fertility control and child health (Raditloaneng 1991). The same is true of this technical sector which directly affects the well-being of women (Joshi 1992). Within countries, Women in Development (WID) is frequently an imported concept rather than a set of policies developed in response to indigenous women's movements and actions. In linking women with welfare, gender is then subject to the analytical and institutional separation which is enshrined in labels like 'women's issues' and 'women's affairs departments' (Lovel and Feuerstein 1985). This marginalisation can be reinforced by a national framework of legal discrimination which further inhibits efforts to promote women in water projects (Kudat and Weidemann 1991).

At the grassroots, in the interests of speed and efficiency, women's participation is frequently appropriated by donors and governments and, on the assumption that their time is freely available, transformed into cheap labour. In practice, it often increases women's already heavy workload (Mills 1992). Domestic constraints including family composition and the development cycle of the household can affect women's ability to participate. For example the availability of healthy adult members to share domestic and productive tasks affects women's time budget. Equally important, when participation involves no change in power relations, improved technology may not necessarily be a benign influence upon their lives (Carr and Sandhu 1988). In seeking to justify women's inclusion as active citizens in these technical interventions, there is also a tendency to resort to profoundly conservative, western-inspired assumptions about gender roles. Apparently 'natural', biological categories—women's nurturing and caring role—are used to target women 'in both water management and hygiene education' (UNDP 1990 p. 2). While this in itself may not be inappropriate, a failure to observe decision-making in particular settings can obscure the importance of men as equally valid targets for health education programmes. Male influence over domestic resource allocation is often crucial in determining whether a water scheme proceeds; health promotion can help to convince them of its worth (Benya 1991; Jongsma 1988).

The tendency to focus on women's role as domestic actors and hence upon their practical as opposed to strategic needs, can obscure the broader range of constraints which shape their lives. Complex social, economic, cultural and environmental processes influence women's ability and willingness to participate (Moser and Peake 1987; van Wijk 1992; World Bank 1989). Evidence suggests that structural adjustment programmes imposed by the World Bank/International Monetary Fund on many African economies, have penalised women in particular through the cuts they require in public services like health and education (Afshar and Dennis 1992). These cuts and their severe social consequences, have been much criticised by UNICEF and other (Jolly 1991). Equally, since women are economic actors in their own right, changes in employment opportunities linked to recession and the effects of drought and flood upon agricultural production, profoundly affect their ability to manage their lives let alone participate in new projects. It can no longer be assumed that women will voluntarily participate in water schemes based solely on the expected health benefits. To many women the availability of credit facilities, support in small-scale enterprise and political representation may be as important as their involvement in social service provision (Berger 1987; Guyer and Peters 1987). Indeed integrated multi-sectoral community water and sanitation programmes have been strongly recommended not only to improve women's living conditions and health but also to enhance their economic well-being through skills training and to increase their earnings through income-generation (Kudat and Weidemann 1991).

The category 'woman' is also cut across by many divisions including ethnicity, class and age. A flawed vision of women's apparent homogeneity fails to recognise how these factors come together to shape the experiences of individual women in particular settings and to influence their access to information and power as decision makers and participants in social infrastructure projects, (Kudat 1988; Moser and Peake 1987; Daswani and D'Cruz 1990). Women are not passive victims of global forces any more than they are passive recipients of development driven by experts. Informal social movements outside government control have a long history of self-help activities. Often antagonistic to the state, they also have experience in lobbying and negotiation to secure from officialdom at least some of the benefits of capital accumulation (Schuurman and van Naerssen 1989). Thus images of universal female oppression should be avoided. Although citizen obligations to assist and contribute are frequently given greater prominence by funding agencies than empowerment and social justice, as vocal political actors women contest official demands for voluntary participation in spheres where they feel 'the state should provide' (Charlton *et al.* 1989; Leahy 1986). Without the benefits of political power, improvements to their lives through welfare, credit or income-generating schemes will be minimal.

From devolution to coordination and centralism

Devolution of power from the centre may be interpreted as a positive attempt to challenge the uniformity of modernity and to promote the work of organisations like NGOs with a proven record of mobilisation at grass roots level. Undoubtedly the new attention directed to them significantly changes their image from organisations good at the grass roots level but a little disorganised and unprofessional in their management and accountability, to agencies at the forefront of the development effort. Collaboration offers a challenge to the superiority of public institutions as agents of progress and to a centralised approach as the most efficient way to reach the largest number of people with improved services (Briscoe and de Ferranti 1988: 20).

'The special role of NGOs and of volunteers' is fully acknowledged given that they are apparently 'flexible, credible, ready and able to experiment in innovative approaches' (UNDP 1990 p.4). Given that donors, governments and NGOs are endlessly varied in their objectives and values, it can be anticipated that so too will be the forms of collaboration (Kardam 1991). They already range from the disbursement of funds from the centre and various subcontracting arrangements to close collaboration in project implementation (Kumar 1991). The World Bank, for example, does not support NGO projects. It is much less flexible in this respect than UNICEF or UNDP. Rather, it continues to fund projects through governments in the hope that NGOs will become involved. Its key role is to give NGOs political respectability within governments. In effect there is no standard model of collaboration.

And what of the impact of these new interactions? In sanctioning NGO activities by incorporating them into officialdom, in effect, seeking consensus through collaboration, new tensions may arise in practice. The 'problem' of social infrastructure and the value of community management is frequently interpreted differently by donors, governments and NGOs. While for officialdom, people's participation may be perceived as a way of working to fight poverty, albeit service provision on the cheap, for many NGOs it is a means of empowering the poor to defeat 'the forces of oppression' variously characterised as a hostile government or the inertia of large bureaucracies (Marsden and Moser 1990). Thus, in formalising collaboration with NGOs, will donor support become a centralised and selective process? In effect, will it become exclusive and excluding? There is indeed evidence of conflict between donors and NGOs, and resistance by NGOs to any kind of material and moral incorporation into donor, and, by implication, government-sanctioned programmes. Tensions can also arise between external NGOs and their indigenous counterparts which, while lacking international credibility, have the benefits of local knowledge and experience. Evidence from the Global Consultation in New Delhi suggests that despite the disparate array of organisations represented, many of whose interests, goals and priorities are at variance with each other, consensus must be reached. These 'other' institutions were to be incorporated into a donor-driven collaborative strategy; in effect, a return to western centralism.

Accountability

Will this institutionalised collaboration impose on NGOs new demands for efficiency and accountability which reduce their effectiveness and flexibility at local level? As donor funds are increasingly disbursed to NGOs, many are adopting the professional and managerial characteristics of private enterprise, albeit non-profit-making and with a social conscience. More positively, how far will collaboration enable NGOs to shape donor/government thinking and practices? Evidence from the Kenya Water for Health Organisation (KWAHO) suggests that this is possible (Narayan-Parker 1988). Governments may be forced to define precisely what a 'supportive and enabling' environment means in practice. But the new progressive image attached to NGOs also needs to be treated with caution. Doubts have been raised over the apparent 'efficiency' and 'people-oriented' nature of many such organisations where alliances with local elites subvert their egalitarian principles (Stevens and Verloen 1985). Equally, as the power of NGOs increases at local level, many governments (including Nepal) now challenge their legitimacy and lack of accountability as service providers. Fearing a loss of autonomy, NGOs in turn, resist the role of intermediary between officialdom and communities (Bratton 1990; Fowler 1991). Thus in practice, a new pattern of local conflict, a mosaic of 'contested places', results.

Commercialism and consumer choice

The task of increasing coverage remains enormous. In 1990 it was estimated that in addition to improving existing services, the population requiring a new service by the year 2000 would be some 813 million in urban, and 1301 million in rural areas (UN 1990). How to achieve these objectives continues to exercise donor agencies. Current thinking seeks to address one major anomaly, namely, a system in which financial subsidy favours the rich. Thus the slogan for the 1990s is 'Some for all rather than more for some' (New Delhi Statement 1991). In order to achieve this, market principles and a commercial emphasis continue to be justified for three main reasons. First, evidence which challenges the grounds for a substantial subsidy to the sector, namely, that only the first 20 litres per person per day constitutes a basic need. Second, the failure of governments to provide the necessary finance to meet unrestrained demand; when subsidised house connections consume large quantities of water there is little left for standpost supplies. Third, a new look at the consumers of services as rational decision-makers.

As part of the process of capacity-building, greater commercial orientation within the public sector is now encouraged to reduce bureaucratic inefficiencies and to promote consumer satisfaction as the primary goal. The case of SANEPAR, a semi-public company providing water services in Parana State, Brazil, provides a successful example (WHO/SIDA 1989). A demand-rather than a supply-driven approach is advocated in which the 'customer' is presented with a range of technologies and levels of service from which a choice can be made based upon willingness to pay (Cairncross 1992). To support this, techniques of social marketing and health promotion are recommended. Small-scale private enterprise is also encouraged which responds to local priorities rather than imposes needs defined outside. The private provision of communal water supply and sanitation has proved to be successful in towns in Ghana. Water retailing from a standpost and, on a smaller scale, an individual with a standpost concession operate effectively in the urban areas of Ghana and Nigeria. Franceys (1993) recommends a 'services supermarket' where potential customers can examine the available technologies and discuss prices, installation services and credit terms. He argues that this approach can work well not only in towns but also in rural areas for discrete technologies such as rainwater catchment tanks, hand dug wells, handpumps and on-plot sanitation. It can also be adapted for communal gravity flow water systems.

This latest approach also raises new questions. How realistic is it to build on the concept of customer power? Admittedly the emphasis on citizen duty and responsibility implicit in community management, is no longer prominent. But in stressing the role of beneficiaries as informed consumers purchasing social infrastructure from the 'best', most cost-effective supplier, donors are assuming that the customer is certain of his/her priorities, has a sound knowledge of the alternatives available and the resources with which to choose. These assumptions need not be accurate. Clean water from a standpost may be a low priority for poor families where the river is nearby and food is scarce (Briscoe and de Ferranti 1988). Equally, consumer choice is a meaningless concept in cases where there is no effective market for services. While low income urban consumers may spend some 20% of household income on water from vendors (well above the 3–5 per cent often quoted as acceptable), in cases where there are no alternatives, this cannot be interpreted as 'willingness to pay'. Insecure housing tenure combined with poverty and indebtedness may also negate the concept of consumer choice where residents are neither able nor willing to take on new debts for water and sanitation improvements (Varley 1987).

Nor can it be assumed that the technology on offer is necessarily the most cost effective way of improving health anyway. Warner and Laugeri (1991) argue that during the Water Decade, undue emphasis by external donors on hardware and technical design—that is, on those aspects of project development that they knew best—reduced the effort given to understanding the complex health linkages or to developing health-related design criteria. In the late 1980s, UNICEF imported over 1 million plastic taps into Nigeria to be attached to water pots. These avoided the need for a contaminated container to be dipped into the household water store. By shifting the focus of intervention from the street to the home, it is anticipated that the health benefits will be much greater than the effort made to 'engineer' ever higher water quality at the standpost. Bradley *et al.* (1992) confirm this point. They emphasise that real health benefits only accrue with yard taps and house connections.

CONCLUSION

The Water Decade was driven by a faith in the benign influence of technology in transforming people's lives. It also gained its authority by drawing on the medical and social sciences. But flaws have appeared in the social and moral concepts of community on which the decade was based. Experience has shown that attractive slogans like voluntary initiative and the role of women, both of which underpin much development work, raise many searching questions. During the course of the Water Decade, donors and governments have found it necessary to clarify and redefine communities and individuals as subjects of change. Clearly active citizenship is about more than changing the community image. While the promotion of self reliance through individual and collective decision making is designed to empower, the sustainability of services is often influenced by conditions unrelated to the technology itself. Drought, changes in employment, population mobility, all influence the ability of individuals to manage their lives, let alone new services.

In the preoccupation with preparing communities for their management role, equally important is how donors and governments respond to the new opportunities and demands. As Yacoob and Warner (1989: 106) point out, 'contrary to the perceptions by many technicians in the field, (community management) is more than an art. It requires professionals, resources and time'. While a top down authoritarian approach requires little more than a sound technical training, philosophies of development which respect cultural and economic diversity confront established professions. A more flexible approach increases the responsibility of, and the adaptability required by, community workers. If, however, officialdom fails to accommodate to these new demands, the danger is that shifting boundaries of difference will arise through a new process of inclusion and exclusion. Communities lacking the modern management skills, those which do not conform to the new 'model community' image or who do not display the necessary consumer individualism, will once again become obstacles to development to be excluded on grounds of their irregular incomes and weak institutions.

During this final decade of the twentieth century, in the continued struggle to combat the 'everyday' morbidity and mortality exacted by endemic diseases like dysentry and diarrhoea, a central challenge is the construction of more imaginative and contingent models which are responsive to the unique experiences and conditions of local environments. It is the variety and complexity of these social and economic landscapes with which 'western-inspired' science, management and technology must come to terms.

REFERENCES

Afshar, H. and Dennis, C. (ed.) (1992). *Women and adjustment policies in the Third World*. Macmillan, London.

Arlosoroff, S. *et al.* (1987). *Community water supply: the handpump option*. World Bank, Washington DC.

Bakhteari, Q. A. (1987). A strategy for the integrated development of squatter settlements: a Karachi case study. Unpublished Ph.D. thesis, Loughborough University.

Bell, M. (1991). Restructuring communities as agents of progress. *Antipode*, **23**, 331–42.

Bell, M. (1992). The water decade valedictory, New Delhi 1990: where pre- and post-modernism met. *Area*, **24**, 82–9.

Benya, P. A. (1991). Community participation in rural water supply and sanitation in Port Lake and Kambia districts, Sierra Leone. Dissertation submitted for the Diploma in Community Water Supply and Sanitation, Loughborough University.

Berger, M. (1987). *A women in development implementation plan for USAID/Bangladesh*. International Center for Research on Women, Washington DC.

Beyer, M. (1988). The human factor. *Decade Watch*, **7**, (1), 1–4.

Biswas, A. K. (1987). Ten years after Mar Del Plata. *Decade Watch*, **6**, (1), 1–3.

Boydell, R. A. (1990). The development of the rural water supply and sanitation sector in Zimbabwe 1974–1987: The design and impact of donor supported projects. Unpublished Ph.D. thesis, Loughborough University.

Bradley, D., Stephens, C., Harpham, T. and Cairncross, S. (1992). A review of environmental health impacts in developing country cities. Urban Management Program Discussion Paper 6, World Bank, Washington DC.

Bratton, M. (1990). Non-governmental organisations in Africa: Can they influence public policy? *Development and Change*, **21**, 87–118.

Briscoe, J. (1992). Poverty and water supply: how to move forward. *Finance and Development*, December, 16–9.

Briscoe, J. and de Ferranti, D. (1988). *Water for rural communities. Helping people help themselves*. World Bank, Washington DC.

Brown, G. A. (1985). To fill the glass. *Decade Watch*, 4, (3), 1.

Buder, S. (1992). *Visionaries and planners. The garden city movement and the modern movement.* Oxford University Press, Oxford.

Cairncross, S. (1992). *Sanitation and water supply: practical lessons from the Decade*. UNDP/World Bank Water and Sanitation Program Discussion Paper 9, World Bank, Washington DC.

Cairncross, S. and Kinnear, J. (1992). Elasticity of demand for water in Khartoum, Sudan. *Social Science and Medicine*, **34**, 183–9.

Carr, M. and Sandhu, R. (1988). *Women, technology and rural productivity: an analysis of the impact of time- and energy-saving technologies on women.* UNIFEM Occasional Paper No. 6, New York.

Charlton, S. E. M., Everett, J. and Staudt, K. (ed.) (1989). *Women, the state and development*. State University of New York Press, Albany.

Christmas, J. and de Rooy, C. (1990). *The Decade and beyond: at a glance*. UNICEF, New York.

Churchill, A. (1987). *Rural water supply and sanitation. Time for a change*. World Bank, Washington D.C.

Cotton, A. and Franceys, R. (1991). *Services for shelter. Infrastructure for urban low-income housing*. Liverpool University Press, Liverpool.

Daniels, D. P. (1992). Participation planning in an era of political transition. The South African experience. Discussion note, 18th WEDC Conference on Water Development and Management, 30 August–3 September, Kathmandu, Nepal.

Daswani, M. and D'Cruz, C. (1990). The right track: A case study of resettlement initiatives by women in Bombay. *Community Development Journal*, **25**, 31–6.

Davies, R. (1990) *Village water committees in the northern region of Lesotho. Options for improving their effectiveness.* Report for the Government of Lesotho under assignment from the Overseas Development Administration, London.

Decade Dossier (undated). New York, UNDP.

Decade Watch (1984). Special 4th Anniversary Supplement, **3**, (3), 5–6.

Dieterich, B. (1983). Whatever happened to the World Water Decade? *Decade Watch*, **2**, (3), 1–2.

Donnelly-Roark, P. (1987). *New participatory frameworks for the design and management of sustainable water supply and sanitation projects.* WASH Technical Report No.52/PROWWESS Report No.50, Washington D.C.

Dossal, M. (1989). Limits of colonial urban planning: a study of mid-nineteenth century Bombay. *International Journal of Urban and Regional Research*, **13**, 19–31.

Drabek, A. (1987). Development alternatives: the challenge for NGOs. *World Development*, **15** (Supplement), ix–xv.

Driver, F. (1988). Moral geographies: social science and the urban environment in mid-nineteenth century England. *Transactions, Institute of British Geographers*, **13**, 275–87.

Elmendorf, M. (1990). *The IDWSSD and women's involvement*. WHO, Geneva.

Esray, S. A., Feachem, R. G. and Hughes, J. M. (1985). Interventions for control of diarrhoeal diseases among young children. *WHO Bulletin*, **63**, (4), 757–72.

Fowler, A. (1991). The role of NGOs in changing state–society relations: perspectives from Eastern and Southern Africa. *Development Policy Review*, **9**, 53–84.

Fowler, A. (1992). Distant obligations: speculations on NGO funding and the global market. *Review of African Political Economy*, **55**, 9–29.

Franceys, R. (1991). Institutional development: 'moving the boundaries'. In: Cotton, A., Franceys, R., Hutton, L. and Pickford, J. (ed.) *WATSAN 2000*. WEDC, Loughborough University, 87–106.

Franceys, R. (1993). Domestic water use: engineering effectiveness and sustainability. In: Natural Resources Institute (ed), *Priorities for water resources allocation and management*. Overseas Development Administration, 43–49.

Glaser, M. (1985). The use of labelling in urban low income housing in the Third World. Case study of Bogota, Colombia. *Development and Change*, **16**, 409–28.

Grover, B. and Howarth, D. (1991). Evolving international collaborative arrangements for water supply and sanitation. *Water International*, **16**, 145–52.

Guyer, J. and Peters, P. (1987). Introduction to special issue on conceptualising the household: issues of theory and policy in Africa. *Development and Change*, **18**, 197–214.

Hartvelt, F. and Okun, D. A. (1991). Capacity building for water resources management. *Water International*, **16**, 176–83.

Harvey, D. (1989). *The condition of postmodernity*. Blackwell, Oxford.

Headrick, D. (1988). *The tentacles of progress: technology transfer in the age of imperialism, 1850–1940*. Oxford University Press, Oxford.

Huston, P. (1985). Taking time. *Decade Watch*, **4**, (2), 1.

INSTRAW (undated). *Women and the International Drinking Water Supply and Sanitation Decade*. INSTRAW, Dominican Republic, Santo Domingo.

Jolly, R. (1991). Adjustment with a human face: A UNICEF record and perspective on the 1980s. *World Development*, **19**, 1807–21.

Jones, J. and Wiggle, I. (1987). The concept and politics of 'integrated community development'. *Community Development Journal*, **22**, 107–19.

Jongsma, D. M. (1988). *The impact of the male primary health care programme on the health knowledge and practices of the population in Afghan refugee villages*. Save the Children Fund, London.

Joshi, E. M. (1992). Women and rural water supply projects. Proceedings of the 18th WEDC Conference on Water, Development and Management, 30 August–3 September, Kathmandu, Nepal, 271–4.

Kalbermatten, J. M. (1983). Nothing short of revolutionary. *Decade Watch*, **2**, (1), 1.

Kalbermatten, J. M. (1991). Water and sanitation for all. Will it become a reality or remain a dream? *Water International*, 16, 121–6.

Kardam, N. (1991). *Bringing women in. Women's issues in international development programs*. Lynne Rienner, Boulder, Colorado.

Kudat, A. (1988). *Communication, women and settlements management*. UNCHS, Nairobi.

Kudat, A. and Weidemann, C. J. (1991). *Gender in water and sanitation sector in South Asia*. Draft document.

Kumar, A. (1991). Delivery and management of basic services to the urban poor: the role of the urban basic services, Delhi. *Community Development Journal*, **26**, 50–60.

Leahy, M. (1986). *Development strategies and the status of women. A comparative study of the United States, Mexico, the Soviet Union and Cuba*. Lynne Rienner, Boulder, Colorado.

Liao, M. and Rogers, L. (1993). A critical analysis of gender strategies in the operation and maintenance of rural water supplies. Paper presented at a conference on sustainable operation and maintenance of rural and urban water supplies in Ghana, 18–23 April, Accra, Ghana.

Lovel, H. and Feuerstein, M. T. (1985). Editorial. Women, poverty and community development in the Third World. *Community Development Journal*, **20**, 156–62.

Lowes, P. (1982). 'Thank you for bringing the river to our houses.' *Decade Watch*, **1**, (2), 1.

MacLeod, R. and M. Lewis (ed) (1988). *Disease, medicine and empire: perspectives on Western medicine and the experience of European expansion*. Routledge, London.

Madeley, J. (1983). 'H' for headache on Mahaweli scheme. *World Water*, June, 34–5.

Marks, S. and Andersson, N. (1988). Typhus and social control: South Africa, 1917–50. In: Macleod, R. and Lewis, M. (ed.) *Disease, medicine and empire*. Routledge, London, 257–83.

Marsden, D. and Moser, C. (1990). Editorial introduction. Participation and housing the poor. *Community Development Journal*, **25**, 3–8.

McCommon, C., Warner, D. and Yohalem, D. (1990). *Community management of rural water supply and sanitation*. UNDP/World Bank, Washington.

Mehretu, A. and Mutambirwa, C. (1992). Time and energy costs of distance in rural life and space of Zimbabwe: case study in the Chiduku Communal Area. *Social Science and Medicine*, **34**, 17–24.

Melchior, S. (1989). *Women, water and sanitation or counting tomatoes as well as pumps*. New York, PROWWESS/UNDP Technical Series.

Mills, C. (1992). *Women, water and sanitation*. Report of a WHO internment, 18 August–6 November, 1992 Geneva.

Moser, C. (1989). Gender planning in the Third World: meeting practical and strategic gender needs. *World Development*, **17**, 1799–825.

Moser, C. O. N. and Peake, L. (ed.) (1987). *Women, human settlements and housing*. London, Tavistock.

Narayan-Parker, D. (1988). *Kenya–people, pumps and agencies. The South Coast handpump project*. New York, PROWWESS/UNDP Technical Series.

Narayan-Parker, D. (1989*a*). *Goals and indicators for integrated water supply and sanitation projects in partnership with people*. PROWWESS/UNDP Technical Series, New York.

Narayan-Parker, D. (1989*b*). *Indonesia—evaluating community management*. PROWWESS/UNDP Technical Series, New York.

New Delhi Statement. (1991). Some for all rather than more for some. *Water International*, **16**, 115–18.

Peterson, J. D. (1991). Squatters in the United States and Latin America: the discourse of community development. *Community Development Journal*, **26**, 28–34.

Qureshi, M. A. (1988). The World Bank and NGOs: new approaches. Paper presented at the Society for International Development Conference on 'Learning from the Grassroots', 22 April, Washington DC.

Raditloaneng, W. N. (1991). Women and men's participation in the water sector: an evaluation study with special emphasis on rural village water supply. Final report to SIDA.

Rathberger, E. (1990). WID, WAD, GAD: trends in research and practice. *The Journal of Developing Areas*, **24**, 489–502.

Report of an Independent Team (1991). A forward-looking assessment of PROWWESS (Promotion of the Role of Women in Water and Environmental Sanitation Services). PROWESS/UNDP, New York.

Rimmer, D. (1981). 'Basic needs' and the origins of the development ethos. *The Journal of Developing Areas*, **15**, 215–38.

Rothermel, T. S. (1990). Letter to delegates, 19 November. UNDP, New York.

Schumacher, E. F. (1974). *Small is beautiful. A study of economics as though people mattered*. Sphere Books, London.

Schuurman, F. and van Naerssen, T. (ed.) (1989). *Urban social movements in the Third World*. Routledge, London.

Sen, G. and Grown, C. (1987). *Development crises and alternative visions*. For development alternatives with women for a new era (DAWN). Earthscan, London.

Stevens, C. and van Verloen, T. (ed.) (1985). *Pressure groups, policies and development. The private sector and EEC—Third World Policy*. Hodder and Stoughton, London.

Stewart, F. (1991). The many faces of adjustment. *World Development*, **19**, 1847–64.

Swanson, M. (1977). The sanitation syndrome: bubonic plague and urban native policy in the Cape Colony, 1900–1909. *Journal of African History*, **18**, 387–410.

Taylor, F. W. (1911). *The principles of scientific management*. Harper and Bros, London.

The Courier (1986). Water Decade: limited results at the half-way mark, No. 96, 64–6.

Timpson, S. L. (1988). People plus: an equation for development (Technical cooperation in the 1990s). Address to the UNDP Senior Staff Seminar, 7 October, Tokyo.

Timpson, S. L. (1989). People-powered development: GO/NGO/donor collaboration in Asia. Keynote Address at the UNDP/FAO/ANGOC national workshop on NGO involvement in grassroots development, 9–11 March, Tagaytay City, Philippines.

Trace, S. and Whiteside, G. (1992). Community participation in project management. Proceedings of the 18th WEDC Conference on Water, Development and Management, 30 August—3 September, Kathmandu, Nepal, 118–20.

UNDP/World Bank (1988). *Water and sanitation: towards equitable and sustainable development. A strategy for the remainder of the Decade and beyond.* World Bank, Washington DC.

UNDP/World Bank (1989). *Community management of water supply and sanitation services.* World Bank, Washington DC.

UNDP (1990). *Global consultation on safe water and sanitation for the 1990s. 'The New Delhi Statement'.* UNDP, New York.

UNDP (1991). *Human development report.* Oxford University Press, Oxford.

UNFPA (1988). Review of operational activities carried out by UNFPA with non-governmental organisations. Unpublished document, UNFPA, New York.

United Nations (1977). *Consolidated action recommendations.* United Nations Water Conference, Mar del Plata, Argentina, 14–25 March.

United Nations (1990). Report A/45/327. Secretary General of the Economic and Social Council to the UN General Assembly, New York.

van Wijk-Sijbesma, C. (1985). *Participation of women in water supply and sanitation. Roles and realities.* Technical Paper 22, IRC, The Hague.

Varley, A. (1987). The relationship between tenure legalisation and housing improvements: evidence from Mexico City. *Development and Change*, **18**, 463–81.

van Wijk, C. (1992). State of the art review. Drinking water supply and sanitation: impacts on women. *Women, Water, Sanitation Annual Abstract Journal*, **2**, 1–5. IRC PROWWESS/UNDF NORAD.

Vaughan, M. (1991). *Curing their ills. Colonial power and African illness.* Polity Press, Cambridge.

Ward, P. (1989). *Poverty, ethnicity and the American city, 1840–1925: changing conceptions of the slum and the ghetto.* Cambridge University Press, Cambridge.

Warner, D. B. and Laugeri, L. (1991). Health for all: the legacy of the Water Decade. *Water International*, **16**, 135–41.

Wisner, B. (1988). *Power and need in Africa. Basic human needs and development policies.* London, Earthscan.

Wood, M. (1992). Community managed water supplies in Ethiopia. Proceedings of the 18th WEDC Conference on Water, Development and Management, 30 August–3 September, Kathmandu, Nepal, 121–4.

World Bank (1988). *List of World Bank financial projects with potential for NGO involvement*. Economic Relations Division, Washington DC.

World Bank (1989). *Women in Pakistan. An economic and social strategy*. PHRWD, Washington DC.

World Health Organisation (1985). *The international drinking water supply and sanitation decade: review of mid-decade progress*. WHO, Geneva.

World Health Organisation (1988). *The international drinking water supply and sanitation decade: towards the targets*. Geneva, WHO/CWS.

WHO/SIDA (1989). People in focus. How the Brazilian water and sanitation company SANEPAR achieved success by putting its staff and customers first. *A human resources development case study No. 5*. WHO/SIDA, Geneva.

Yacoob, M. (1984). Identification and use of grassroots community organisations in low cost urban housing: the case of Ahmadiyya-type organisations in the Ivory Coast. *African Urban Studies*, **18**, 49–58.

Yacoob, M. (1988). Turning things around: community participation after the fact. *Water Quality Bulletin*, **13**, 136–7.

Yacoob, M. and Warner, D. (1989). Community participation in water and sanitation projects. *Water Quality Bulletin*, **14**, 105–6.

Yeoh, B. S. A. (1992). Municipal sanitary ideology and the control of the urban environment in colonial Singapore. In: Baker, A. R. H. and Biger, G. (ed.) *Ideology and landscape in historical perspective. Essays on the meanings of some places in the past*, p. 148–72. Cambridge University Press, Cambridge.

Yudelman, S. (1987). The integration of women into development projects: observations on the NGO experience in general and in Latin America in particular. *World Development*, **15** (Supplement), 179–89.

5

Development, Modernisation and Health Intervention

STANLEY J. ULIJASZEK

INTRODUCTION

Health is a logical indicator of development, since it reflects improved well-being of individuals and populations; however, most economists do not use it. In this chapter, the relationship between health, development, modernisation and intervention are examined, starting with a brief overview of what intervention, development and modernisation are, and how development is measured. This is followed by descriptions of development models for health improvement. Subsequently, the health impacts of development, modernisation and intervention are examined in two settings in Papua New Guinea, one in the Highlands, the other, the Highlands fringe. This analysis is then extended to consider to what extent health changes have been due to economic development and the process of modernisation which has accompanied it, and to what extent they may be attributable to specific health interventions such as primary health care and immunisation.

It is important to differentiate development and modernisation from intervention. Interventions are targeted changes to the biosocial environment aimed at improvement in human well-being, health or welfare. The means by which such changes are introduced are various, and may be either direct, or indirect. Development includes the idea of economic improvement and the social and cultural changes which accompany it, and is thus broader in scope than intervention. Development is linked to theories of international relations which traditionally placed societies on an evolutionary scale, with the western industrialised nations being considered most advanced, or developed. In this scheme, the nations not yet having undergone the transformations toward prosperity and economic growth are labelled undeveloped, underdeveloped, or less developed, according to their stage of economic transition. However, the development concept over-simplifies a highly complex series of variables, including the traditional economic, political, social and cultural characteristics of different groups and populations, and the ways in which they might change in response to the introduction of new technologies and ideologies. Relationships between developed and developing countries clearly exist, it is conventional to consider the process of development within the framework of a capitalist world system of colonial and neo-colonial dominance (Seymour-Smith 1986).

It has often been assumed that hypotheses formulated to account for the process of industrialisation in Western nations may also apply to the development in other parts of the world (Horowitz 1972). Attempts to delineate the necessary conditions for industrialisation have focused on various factors in creating socio-economic change, including production, labour quality and personal psychological make-up; psychological traits such as the desire for achievement, and the motivation to change have been linked to the entrepreneurial role in societies where the achievement oriented personality is common. However, such psychological factors cannot be considered causal of social change, since social and economic forces may be favouring or shaping the personality for achievement.

In the developed world, the term modernisation encompasses all the developments which followed in the wake of industrialisation and mechanisation, and include the loosening of boundaries between social classes, increased social mobility, as well as the growth of widespread education, the evolution of procedures of industrial negotiation, and the development of social welfare systems at the national and regional level. It also includes processes such as the rise of bureaucratic states, the spread of capitalism, and the secularisation of culture and education (Smith 1994). These things have happened in most western nations, as their populations underwent the transition from a predominantly agricultural mode of subsistence, to one of wage-earning within an industrialised cash-based economy.

The experiences of the newer nation states becoming independent in the twentieth century are quite different from those of western nations which underwent their industrial revolutions in the eighteenth century. In this context, modernisation has been defined as the interaction of less complex energetic, technological and socio-economic systems characterized by regional production and consumption with contemporary economic systems of industrial technology influenced by the national and international market, social and political factors (McGarvey 1989). This acknowledges the framework of a capitalist world system of colonial and neo-colonial dominance, but it does not suggest which aspects of change that arise in the course of such interaction lead to changes in health.

MODELS OF HEALTH IMPROVEMENT

An underlying assumption about health in relation to development is that increased material well-being will result in increased physical well-being. Material improvements, along with associated social changes, have been associated with reduced fertility and mortality in many parts of the world (Caldwell 1982; Ruzicka and Kane 1990).

Development has been described as a comprehensive process (Hammarskjold Foundation 1976) informed by five principles: 1) the satisfaction of human needs, both material, and in a broad sense, political; 2) self-reliance; 3) endogenous growth; 4) harmony with the environment; and 5) structural transformation in the economy, society and polity. This suggests a scheme of human activity that involves constant transformation of human and environmental structures toward material benefit, with checks and balances that prevent both the destruction of natural resources and degradation of the environment and the creation and maintenance of social and material inequalities. In places where inequality may have existed, it can be perpetuated, or even amplified. Furthermore, unless the traditional structures incorporated notions of environmental conservation, it may be impossible to achieve harmony with the environment.

The problem of assessing the success or impact of development arises in relation to changing regional and traditional economic structures and systems as part of the usually dominant national and global economic system. Economists use markers, such as gross national product (GNP), gross domestic product (GDP), and economic growth per annum, to assess the success of the development process in different nations. At more local and regional levels, measures of changing income per capita, proportion of people involved in the cash economy, and trade involving cash transactions are used to assess the relative welfare of different communities and populations. However, there is often a wide empirical and conceptual gulf between capitalist and precapitalist economic formations, and the measures employed by economists may be of limited use in approaching the measurement of the true impact of economic development. In this regard, the concept of modernisation is useful, since it aims to describe all the changes that take place in relation to involvement in the global economic system, whether it arises out of industrialisation and urbanisation, or cash-cropping and changing subsistence orientation in rural areas. Economists generally dismiss the use of health and demographic indicators in the assessment of development, even though an underlying assumption of this process is that economic growth will result in an improvement in the material conditions of populations, with an associated improvement in well-being. Although primary health specialists and development planners are aware that measures such as mortality, life expectancy, food availability and nutritional status can give a more direct measure of the well-being of populations, the relationships between economic indicators, and health and demographic indicators are not simple.

Economic change and undernutrition

Per capita GNP is a widely used index of relative economic status of different countries, with change in GNP across time being used to determine economic progress within any nation. However, comparisons of national-level figures for GNP, and GNP change with measures of well-being, such as nutritional status, per capita food availability, infant mortality rate (IMR), or life expectancy at birth (LEB) do not give clear-cut relationships.

Nutritional indicators

Nutritional assessment is a means by which the health status of individuals or populations as influenced by their intake and utilisation of nutrients can be determined. This involves the interpretation of information obtained from any of dietary, biochemical, anthropometric or clinical studies (Gibson 1990) and is often not easy to interpret. Anthropometric measures include height for age and weight for age, relative to growth references such as those produced by the National Center for Health Statistics (1977). The proportion of the young child population below some cut-off of these references is used as a measure of undernutrition-induced stunting and wasting, respectively. Although one anthropometric measure of current nutritional status is low weight for height relative to western references (Waterlow *et al.* 1977), the most widely used index in the developing world is low weight for age, which reflects both past and present exposure to the dual stresses of undernutrition and infection.

A simple but objective measure of current nutritional status in adults is the body mass index (BMI), or weight (kg) divided by height (m) squared. It is closely related to food consumption, is a sensitive index of physical function and performance, is sensitive to socio-economic status and has been recommended for use in regional and national surveys for the purposes of nutritional monitoring and surveillance (Shetty and James 1994). It has also been proposed for the evaluation of nutrition interventions (Shetty and James 1994). There are problems with its use in that this weight-height relationship does not necessarily reflect body composition similarly in different populations (Norgan 1990), although it correlates well with body fat composition within-populations (Shetty and James 1994), at least in Europeans. BMI below 18.5 has been proposed as the cut-off for undernutrition internationally, with lower cut-offs delineating greater extremes of undernutrition (James *et al.* 1988; Ferro-Luzzi *et al.* 1992). Possible relationships between BMI and economic indicators have yet to be explored.

The most commonly used method for assessing the nationally available food supply is based on food balance sheets. These measure the food which is available for consumption, but not necessarily that which is consumed. Food balance sheets are a national account of the annual production of food, changes in stocks, imports and exports, and distribution of food over various uses within the country (Food and Agriculture Organisation of the United Nations 1980). Although they are more accurate than measures of individual food intake (Ulijaszek 1992), they cannot be used to determine food consumption in relation to regional, economic, demographic, seasonal, or socio-economic differences within a country (Gibson 1990). Their use has little to do with nutritional assessment. Rather, they provide information for food and agricultural policy at national level, and allow international comparisons of food availability. Thus they are markers of dietary well-being, and can be used in the context of development.

Demographic measures

Infant mortality rate is the number of children dying in the first year of life, per thousand live births per year. It includes deaths during childbirth, and therefore reflects maternal factors, as well as susceptibility factors related to the maternal environment, and direct exposure to the disease and nutritional environment. Susceptibility factors include nutritional and immunity factors *in utero* and while breastfeeding, and the nature of maternal mediation between the infant and the environment, including the duration of breastfeeding or the decision not to breastfeed, the nature and timing of dietary supplementation to the infant, and sickness behaviours and management of the infant when ill. Thus it is a composite measure, which to some extent is similar to weight for age as a measure of the effects of undernutrition and infection on child growth, but which gives an estimate of the most extreme negative health outcome, death. Infant mortality is given particular attention above mortality estimates in other age groups, since this age group has the highest mortality rate in developing countries. Although this is usually reported at national level, it is often available at regional and community level.

Life expectancy at birth (LEB) reflects the mortality experience across the lifetime of a population, either at national, regional or community level. It is the number of years new-born children would live if subject to the mortality risks prevailing for the cross-section of the population at the time of their birth (United Nations Children's Fund 1990). It is a function of infant and young child mortality, and in populations where this is high, LEB is low. However, it provides an integrated measure of subsequent quality of life, measured by the most extreme outcome of poor quality of life, which is death.

Nutritional indicators and development

Table 5.1 gives the extent of nutritional wasting (children aged 0 to 5 years below 80% of National Center for Health Statistics (1977) fiftieth centile of weight for age) in Asia, Africa, the Middle East and North Africa, and in Central and Latin America, in the period 1980–1987 (United Nations Children's Fund 1990), while Fig 5.1 and 5.2 show relationships between the extent of nutritional wasting, and GNP in 1987, and GNP change between 1980 and 1987 respectively, in the same countries. These data are far from complete, since no

TABLE 5.1 The extent of nutritional wasting in different parts of the world, 1988 (from United Nations Children's Fund 1990)

	%	millions
ASIA		
Overall:	43	
Bangladesh	60	11.1
Vietnam	52	4.8
Indonesia	51	10.9
India	41	46.1
Pakistan	39	8.7
Myanmar	38	2.1
Sri Lanka	38	0.7
Laos	37	0.2
Philippines	33	3.0
Thailand	26	1.6
Kampuchea	20	0.3
AFRICA		
Overall:	34	
Mozambique	57	1.5
Niger	49	0.6
Tanzania	48	2.5
Mauritania	44	0.1
Sudan	41	1.8
Ivory Coast	40	1.0
Ethiopia	38	2.9
Burundi	38	0.3
Rwanda	37	0.5
Ghana	37	1.0
Liberia	35	0.1
Benin	34	0.3
Madagascar	33	0.7
Mali	31	0.5
Central African Republic	30	0.1
Zambia	28	0.4
Sierra Leone	23	0.2
Malawi	22	0.3
Senegal	22	0.3
Zaire	20	1.3
Cameroon	17	0.3
Lesotho	16	0.05
Botswana	15	0.03
Zimbabwe	12	0.2
MIDDLE EAST AND NORTH AFRICA		
Overall:	24	
Yemen Arab Republic	61	0.8

TABLE 5.1 *Contd.*

	%	millions
Iran	43	4.1
Yemen Democratic Republic	26	0.1
Syria	25	0.6
Morocco	16	0.6
Turkey	12	0.8
Egypt	11	0.9
Kuwait	6	0.02
CENTRAL AND LATIN AMERICA		
Overall:	14	
El Salvador	55	0.4
Guatamala	34	0.5
Paraguay	32	0.2
Panama	16	0.05
Bolivia	15	0.2
Peru	13	0.4
Brazil	13	2.4
Colombia	12	0.5
Nicaragua	11	0.1
Equador	10	0.2
Venezuela	10	0.3
Uruguay	7	0.02
Chile	3	0.04

national figures are available for the extent of undernutrition in a number of countries, including China, Nigeria, Kenya, South Africa, Mexico, Argentina and the Libyan Arab Jamahirya. Although the methods of ascertainment and the degree of reporting vary from country to country, making the data on nutritional status less than ideal, some interesting patterns and relationships are revealed. Overall, about 38% of children under five years of age in developing countries are undernourished, the greatest proportion being found in Asia (43%), the lowest in Central and Latin America (14%). However, the value for Asia is likely to be an over-estimate, since data for the more modernised nations such as Malaysia, Taiwan, Japan and South Korea are not available, and it cannot be assumed that there is no undernutrition in those countries.

For the Asian countries reported, Bangladesh has the highest proportion of the under-fives population undernourished, followed by Vietnam and Indonesia. Although a lower proportion of under-fives is undernourished in India, the sheer size of the Indian population gives that country by far the greatest number of undernourished

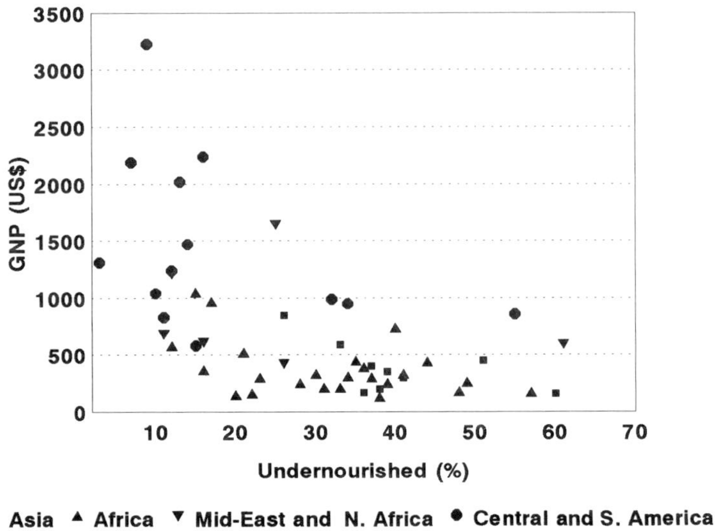

Fig. 5.1. The relationship between per capita gross national product (GNP, US$) and the proportion of children aged 0–5 years below 80% of the National Center for Health Statistics (1977) fiftieth centile of weight for age in developing countries of Asia, Africa, the Middle East and Central and South America, 1988. (Data from United Nations Children's Fund 1990)

for the Asian countries for which national data is available. There is surprisingly little undernutrition in Laos and Kampuchea. Of the African countries, Mozambique, Niger and Tanzania have the greatest proportions of their under-fives populations undernourished, while Ethiopia, Tanzania and Sudan have the largest absolute numbers of undernourished children. The countries with the lowest prevalences of undernutrition are Lesotho and Malawi, on the basis of national data collected between 1980 and 1987.

In the Middle East and North Africa, the Yemen Arab Republic and Iran have the greatest proportions of undernourished children; Iran also has the greatest total number of undernourished. No data is available for Iraq. In Central and Latin America, El Salvador has the highest proportion of undernourished children, while Brazil has by far the largest total number. Uruguay and Chile have both the lowest proportions of undernourished children, and the lowest total numbers.

Figure 5.1 shows that the relationship between GNP and undernutrition at the international level is weak, although clearly evident, and curvilinear. However, at lower levels of economic prosperity, GNP does not predict the extent of undernutrition to any great degree. For example the range of nutritional wasting prevalence in nations with GNP around US$1000 is 10–55% of all children below 5 years of age, while in nations with GNP below US$500, the range is 16–60%. However, within-country, the relationship can be quite clear. Table 5.2 shows the proportion of preschool children classified as malnourished on the basis of weight for height and/or height for age below − 2 Z scores of the NCHS (1977) growth references, in Panama (Franklin *et al.* 1985). In general, a greater proportion of the offspring of members of occupation groups with incomes below the national average are classified as malnourished, although some groups have anomolously high or low levels of malnutrition in relation to parental income group. For example the children of diversified small farmer owners have the greatest rate of undernutrition although their income is slightly above the national average; conversely, self-employed workers, whose income is below the national average, have one of the lowest rates of malnutrition. This would indicate that there are other factors which mediate in the

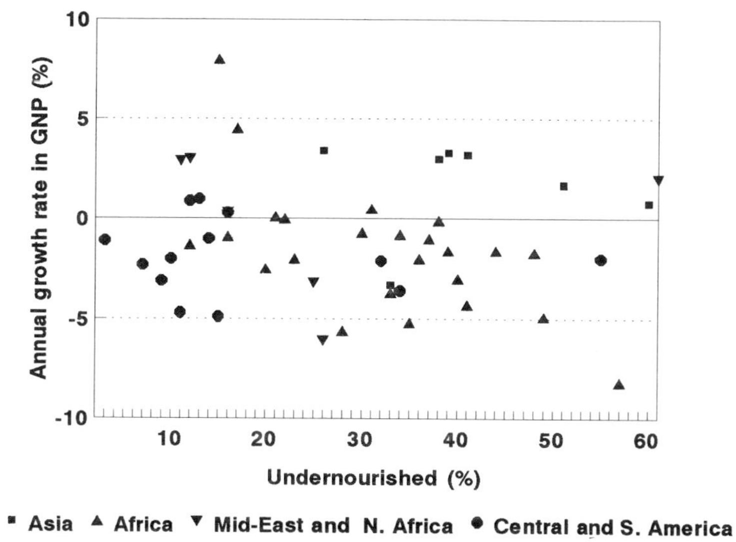

Fig. 5.2. The relationship between annual growth rate in per capita gross national product, 1980–87, and the proportion of children aged 0–5 years below 80% of the National Center for Health Statistics (1977) fiftieth centile of weight for age in developing countries of Asia, Africa, the Middle East and Central and South America, 1988. (Data from United Nations Children's Fund 1990)

TABLE 5.2. Income and preschool malnutrition in Panama (from Franklin *et al.* 1985)

Occupational group	Income (US$ per annum)	Undernourished preschoolers[a] Sample size	%
Commercial transport workers	288	46	11
Professionals and office workers	198	45	11
Salaried urban workers	150	42	16
Skilled workers	150	81	16
Farmers on diversified small farms	127	213	30
Agricultural workers	114	154	21
Farmers with employees	113	40	23
Unskilled workers	96	92	21
Horticulturalists	87	157	25
Self-employed workers	80	35	13
Maize producers	78	26	20
Rice producers	66	38	24
Producers of maize and rice	56	84	25
National average	122		20

[a] Total of children below – 2 Z scores of either or both of National Center for Health Statistics (1977) references of weight for height or height for age

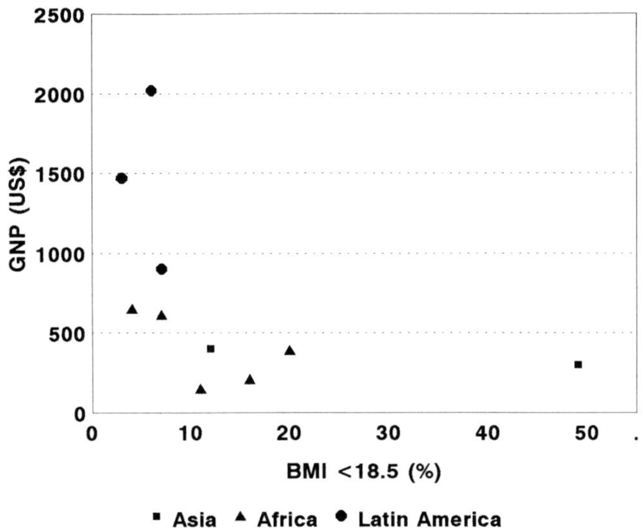

Fig. 5.3. The relationship between per capita gross national product and the proportion of adults undernourished (body mass index below 18.5) in ten developing countries. (Data from United Nations Children's Fund 1990; Shetty and James 1994)

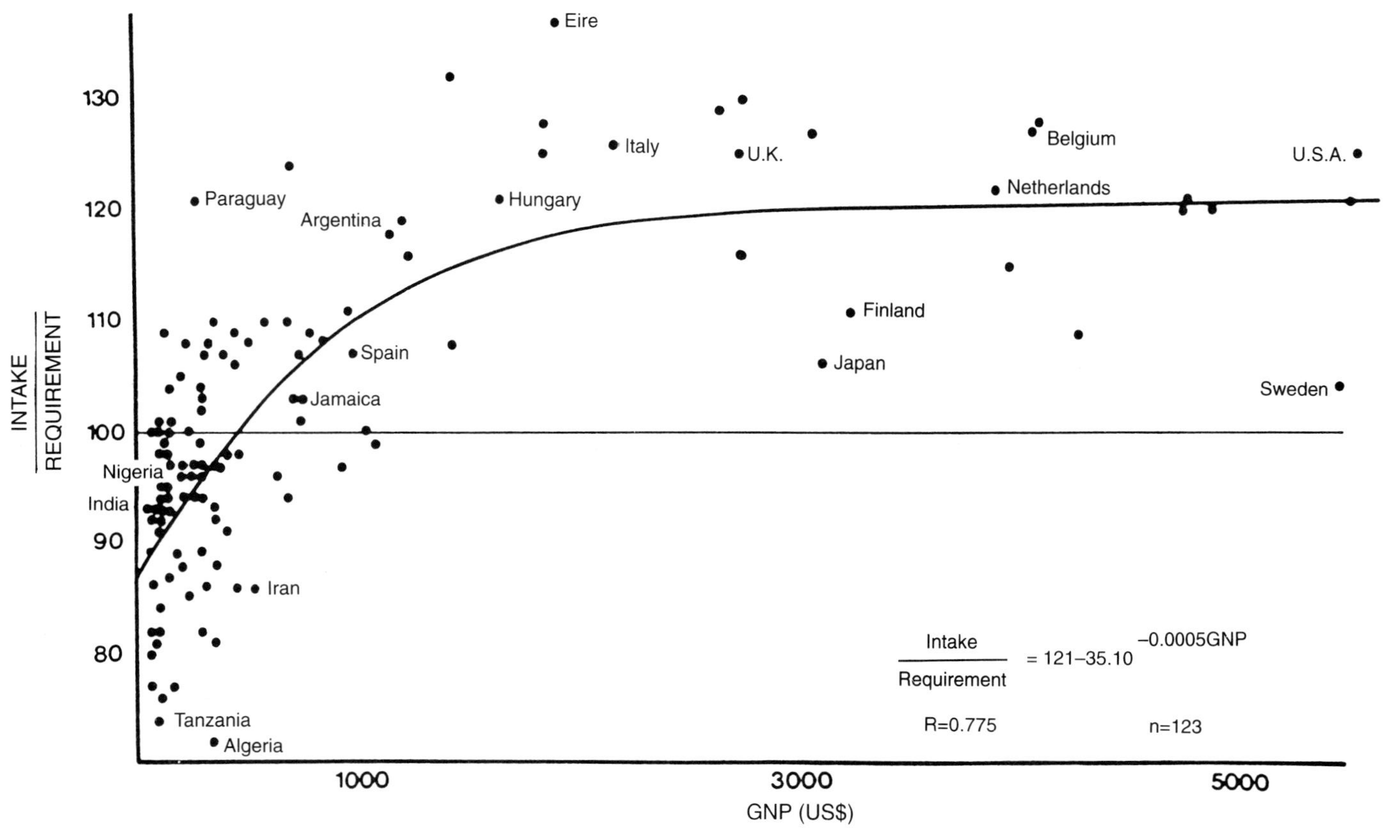

Fig. 5.4. The relationship between per capita gross national product (Organisation for Economic Cooperation and Development 1974) and energy intake, as food moving into consumption, expressed as a percentage of requirement (Food and Agriculture Organisation 1974). (From Miller 1979)

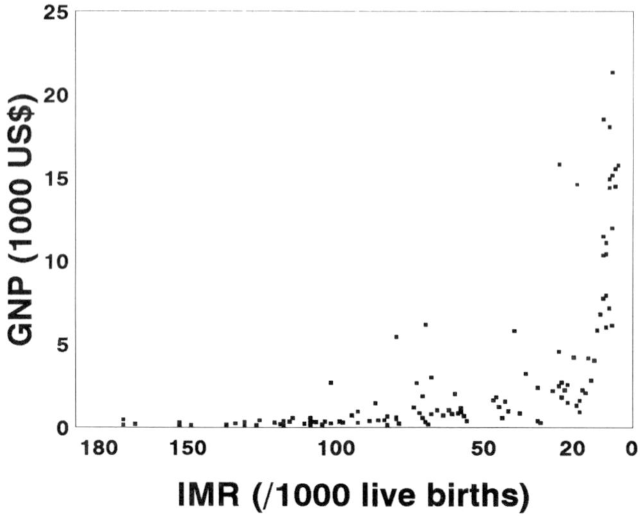

Fig. 5.5. The relationship between infant mortality rate in 1988 and per capita gross national product in 1987 across 120 nations. (Data from United Nations Children's Fund 1990)

effects of the nutritional ecology on nutritional status of children. In general, it might be assumed that income and educational level are associated, and that in part the nutritional influence of income is mediated through the educational level of the parents, inasmuch as behaviours involved in within-household food allocation and illness management can have an impact on child nutritional status and growth patterns. However, the achievement motivation of self-employed workers might outstrip their current earning capacity, and might also be associated with progressive behaviours which could ameliorate the extent of undernutrition in that group.

There is a weak relationship between economic growth and the extent of undernutrition, with those countries with the fastest rates of increase in GNP having lower levels of undernutrition (Fig 5.2). There is also a relationship between GNP and extent of undernutrition in adults. Figure 5.3 shows that in ten populations for which the proportion of adults with body mass index (BMI) below 18.5 varies between 3% (Peru) and 49% (India), there is substantially more adult undernutrition in nations with GNP below US$500.

The case that development improves food supply has been argued by considering relationships between economic indicators and per

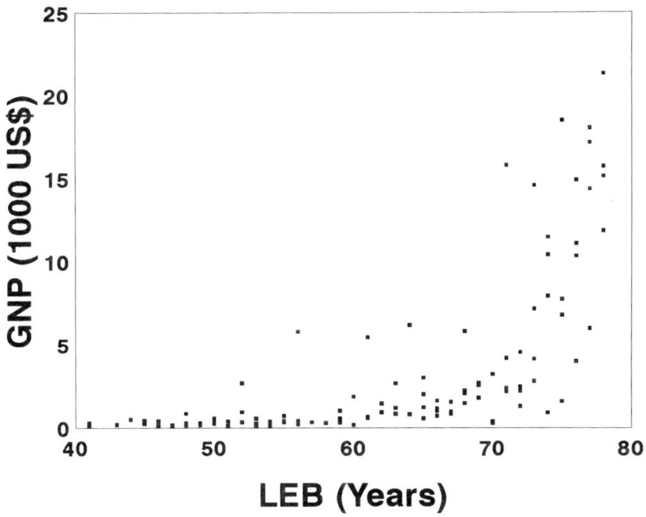

Fig. 5.6. The relationship between life expectancy at birth in 1988 and per capita gross national product in 1987 across 120 nations. (Data from United Nations Children's Fund 1990)

capita food availability. Figure 5.4 shows GNP and mean per capita food availability expressed as a percentage of FAO/WHO (1973) requirement for 123 nations. The relationship is curvilinear, in which small differences in national economic status are associated with big differences in food availability at the lowest levels of national income, but no differences at the highest levels of income. Although this has been taken to show the importance of economic development to nutritional well-being at national level, such relationships are simplistic. Apart from the likely social disparities at the regional, local and community level which can arise from the process of modernisation, the relationship between national income and other indicators of health such as IMR and LEB may differ according to GNP. Figures 5.5 and 5.6 show the relationships between IMR and GNP, and LEB and GNP respectively, for 120 nations (from the richest to the poorest) in 1988.

There is no great difference in GNP between nations that have IMRs above about 20/1000 live births. However, there is an enormous spread of IMR in nations with GNP below $500 per capita, ranging from 172/1000 (Angola and Mozambique) to 31/1000 (China). This suggests that the very high IMRs (180–100/1000 live births) in some developing countries may have little to do with the absolute level of economic development. Indeed, reductions in IMR from about 180/1000 to about 100/1000 might not be best achieved through economic development. Rather, low-cost health interventions might be a more effective way of attempting health improvement. In countries with IMRs below this level, economic development might be more effective. However, reducing IMRs below about 20/1000 may be an impracticable goal in the absence of highly technologically-based medical services which have a broad outreach. At present, conditions for such services are to be found only in the industrialized nations of the West and the former Eastern European bloc, and rapidly modernised and modernising countries such as Japan, Taiwan and Korea. A major challenge for health planners in many developing countries is the reduction of IMR from 50–60/1000 live births to around 20/1000, without substantial increases in health expenditure.

Among the poorest nations, there is little difference in GNP, but a very wide range of life expectancy at birth (Fig. 5.6). For nations with GNP below $500 per capita per year there is a spread in LEB from 41 years (Sierra Leone and Ethiopia) to 70 years (China, Guyana and Sri Lanka). Increasing LEB, from the lowest levels reported, to about sixty years, might be better achieved by means other than economic development. However, increasing life expectancies beyond this probably requires more than health intervention.

Cash cropping

The commercialisation of agriculture has been seen as an important element in efforts to increase incomes and improve living standards in rural areas of developing countries. Braun and Kennedy (1986) have examined the competition between cash crop and food crop production and its effects on income and nutrition. By correlating various cash cropping variables with staple food production for 78 developing countries, they found associations between: 1) increased area of land devoted to cash cropping and increased staple food production ($r = 0.32$, $p < 0.01$); 2) increased proportion of land devoted to cash cropping and per capita income ($r = 0.33$, $p < 0.01$) and increase in per capita income ($r = 0.26$, $p < 0.05$); and 3) increased proportion of land devoted to cash cropping and reduced per capita staple food production ($r = -0.27$, $p < 0.05$). Thus it would seem that the greater the wealth, the greater the adoption of cash cropping. Furthermore, the greater the adoption of cash cropping, the greater the increase in income and in overall staple crop production, but the lower the per capita staple food production. Braun and Kennedy do not present demographic data in their analysis, but this discrepancy may be due to increased overall food requirements as a result of population growth, which are not matched by production.

Within-countries, the relationships between cash cropping and staple food production and income are varied (Table 5.3). In most countries, increases in the area of land devoted to cash cropping is associated with the level of staple food production. However, unlike the between-country comparison, increased proportion of land devoted to cash cropping is positively associated with per capita staple food production in more countries than it is not. The countries with a negative association between these two variables are most likely to be those with larger proportions of the rural population engaged in wage labour, rather than, as Braun and Kennedy suggest, to reflect the failure of some countries to manage both cash and staple food cropping. Furthermore, within-country comparisons show that increasing the extent of cash cropping has no association with income per capita in most countries, and a negative association in most of the others. Many of the 22 countries with a negative association between these two variables are those with deteriorating economic growth but

TABLE 5.3. Within-country associations between cash cropping and staple food production and income in various developing countries, 1968–82 (from Braun and Kennedy 1986)

Patterns and changes in cash cropping	Country-specific basic data	Correlation coefficient[(a)]; number of countries with associations that are: Positive	Negative	Not significant
Area under cash crops	Basic staple food production	44	4	30
Cash cropping as a proportion of total crop land used	Per capita basic staple food production	24	15	39
Cash cropping as a proportion of total crop land used	Per capita gross national product	2	22	54

[(a)] Correlation coefficients significant at $p < 0.1$ are used for classification into the positive or negative group

TABLE 5.4. Effects of cash cropping on food intake and/or nutritional status

Country	Cash crops	Effect on food intake/nutritional status	Reference
Brazil	Sisal	– ve	Gross and Underwood 1971
Mexico	Cocoa, sugar cane	– ve	Hernadez *et al.* *1974*
Mexico	Cocoa, sugar cane	– ve	Dewey 1979
Tanzania	Coffee, bananas	+ ve	Lev 1981
Kenya	Tea	+ ve	Hitchings 1982
Kenya	Coffee	+ ve	Hitchings 1982
Kenya	Cotton	0	Hitchings 1982
Kenya	Pyrethrum	0	Hitchings 1982
Kenya	Sugar cane	– ve	Hitchings 1982
Kenya	Coffee, staple foods	+ ve	Rabeneck 1982
Kenya	Coffee, vegetables	+ ve	Fleuret and Fleuret 1980
Papua New Guinea	Coffee	+ ve	Harvey and Heywood 1983*a*
Kenya	Sugar cane	0	Kennedy and Coghill 1987
Kenya	Rice	– ve	Niemeyer *et al.* 1988
Swaziland	Cotton	– ve	Huss-Ashmore and Curry 1989
Cameroon	Cocoa, food	O	Guyer 1989
India	Rice	+ve	Pinstrup-Anderson and Jaramillo 1989

with a relatively stable allocation of land to cash crops (Braun and Kennedy 1986). Thus it would seem that in most countries, increased cash cropping has a positive influence on staple food production, but no real effect on income.

The influence of cash crop production on nutritional status has been examined by a number of authors, and these are summarised in Table 5.4. Although it has been suggested that the introduction or extension of cash crop production may adversely affect nutritional status (Dewey 1979; Fleuret and Fleuret 1980), there is no general trend. Rather, it may depend on the type of cash crop planted; coffee growing seems to have a positive effect on nutritional status, while sugar cane and cocoa cash cropping have a negative effect. This may relate to the price of these crops on the international market, and the nature of work organisation and pay of wage labourers engaged in this type of work. In some places, coffee growing is carried out on a smallholder basis, and this may increase the relative financial return relative to work exerted in cash cropping. In some places, surplus labour availability may drive down wages such that workers engaged in labour for cash crop plantation owners may be financially exploited in a way that their nutritional status suffers. However, the nutritional effects of commercial farming may be specific to time and place, and are more likely to be a function of total economic strategies, including crop production and other sources of income (Huss-Ashmore 1993). For example in rural Mexican households, child nutritional status was negatively related to reliance on sporadic migrant wages and seasonal illness, but not to the proportion of income from cash cropping (Kaiser and Dewey 1991).

Thus it would seem that relationships between indices of economic development and of well-being as assessed by nutritional measures, food availability and population indices of mortality and longevity are at best weak. Furthermore, generalisations about the nutritional effects of cash cropping have been shown not to hold true. Modernisation studies are more likely to tease out the relationships between economic change, material improvement, and health and well-being, since they examine a greater range of variables which measure aspects of all these factors.

Modernisation

Modernisation includes all of the broader social and economic outcomes of the development process. These are difficult to understand and predict, since socio-economic and local population factors, including income, wealth, land distribution, land tenure systems, proximity to urban centres, population density and changes in subsistence practices, interact with the increased exposure to contemporary national and international economic forces to produce a range of outcomes from relative and absolute impoverishment, to varying degrees of enrichment (McGarvey 1992). Furthermore, socio-economic factors operate within groups and local populations, and development can serve to increase or amplify any social and economic disparities that might exist. Development may also create disparities where they did not previously exist, by allowing differential access to new resources. Once established, such disparities are likely to be perpetuated. Modernisation studies examine biosocial relationships between social and economic constraints to resources and technologies, changes and amplification of power and status relationships, cultural incongruities created by change, and the health and biological outcomes of such changes.

Such work has been carried out in relation to health and biological outcomes in Samoa (McGarvey and Baker 1979; Bindon and Baker 1985; McGarvey *et al.* 1989), New Guinea (Ulijaszek *et al.* 1987), among Amerindian populations (Szathmery 1990) and Australian Aboriginal people (O'Dea *et al.* 1980, 1982). In these, good measures of socio-economic status and development are often lacking, and linkages between economic development, the broader effects of modernisation, and health outcomes have been difficult to examine as a consequence.

ECONOMIC DEVELOPMENT AND MODERNISATION IN PAPUA NEW GUINEA

The process of economic development, modernization and change in health and well-being of the populations of one geographical region, East New Guinea (which in 1975 became independent Papua New Guinea) is described in some detail here. Various New Guinea populations are at different stages of modernisation and economic development, and it is possible to relate economic change and modernisation to health outcomes in a historical context, and in respect of rural development, cash cropping, mining and urbanisation.

The colonisation and early economic development of New Guinea

The name 'New Guinea' was coined in 1545, when the Spanish explorer Ynigo Oritz de Retes formally claimed Western New Guinea (currently Indonesian Irian Jaya) in the name of the King of Spain (Tudor 1969). There had been both local trade and trade with more distant populations in New Guinea since prehistory, and with small numbers of European merchants in the seventeenth century, but the first significant contact with an external dominant culture came with the claim in 1793 of parts New Guinea as a sphere of interest for the British East India Company (Tudor 1969). This had little real impact on the local economies, since most of the attention paid to this land mass was in relation to exploration and the mapping of territory. The country was first annexed for Britain in 1846, by Lieutenant Yule of the Royal Navy, but he was ignored by his government. The New Guinea island was colonised during the scramble for colonies among the South Pacific Islands which ensued after the Franco-Prussian War of 1877, when Germany suddenly became a colonising power. The Germans chose to acquire the north-western quarter of the New Guinea mainland, as well as several large islands to the east of this. This alarmed authorities in Australia and New Zealand who petitioned Britain to annex what it could of New Guinea, and political events around 1880 resulted in the creation of German New Guinea and British New Guinea (Ulijaszek 1993). Formal annexations of British and German New Guinea took place in 1884 and 1885 respectively (Souter 1961). In 1899, an agreement between the United States, Britain and Germany left Bougainville and Buka in German hands, and they were incorporated as part of German New Guinea.

Between 1884 and 1906, exploration, and the establishment of government influence resulted in the discovery of goldfields in Misima and Woodlark Islands (of the Milne Bay Islands), Gira, on the Mambare River, and Yodda, in the Papuan interior. However, much of the initial concern was with plantation development, under the administration of white settlers. The first plantation crops to be introduced were the cash crops copra, rubber, sisal and coffee (Dick and McKillop 1976), all cash crops.

Initially this had very little effect on the traditional New Guinea economies, apart from the cash income now available to indentured labourers working in plantations, either in New Guinea or elsewhere. Recruitment of New Guineans for work on plantations in other Pacific areas including Queensland and Samoa started almost simultaneously with external trade (Epstein 1972). Young men were often reluctant to volunteer for labour migration, and in some places, arrangements involving the payment of indigenous big men with steel tools, guns and other introduced goods for a regular supply of labourers were established by middle men. Although the immediate effect of this external trade on the traditional economy was a more efficient use of available resources, its extent was still rather limited. This began to change in 1895. In British New Guinea, the Lieutenant-Governor Sir William MacGregor (1895) outlined a policy on economic development for Papuans. He believed that the well-being of Papuans would improve from their present dire state if they were introduced to the practice of cash cropping at the village level. This was implemented by legal compulsion in British Papua in 1894, and a similar policy was pursued by the authorities in German New Guinea soon after (Ulijaszek 1993). Despite criticism from European planters, who felt that opportunities for earning independent cash incomes would discourage indentured labour, government officials actively encouraged villagers to grow coconuts and vegetables for the urban market, and to collect rubber. Thus, almost from the outset of colonial administration, the aim was to integrate native society into the cash economy.

By the early twentieth century, the modernisation process was established in the coastal regions, with the instigation of both plantation and smallholder based cash cropping. In addition, the creation of colonial administrative centres saw the beginnings of urbanization as paid labour became available in the new townships. With this came the development of market gardening for the urban market, and the introduction of imported foods which could be bought with cash. In addition, the provision of medical services expanded from less than 2% of total administration expenditure in 1900–1, to over 8% in 1910 (Gunther 1972), a level that has been maintained ever since. In 1990, the proportion of gross national product spent on health care was 8.1% (Policy Planning and Research 1991).

Plans for the development of native agriculture were drawn up by the Department of Agriculture in 1924, and a Division of Agricultural Education was created in 1927. Prior to the Second World War attempts to integrate the rural population into the cash economy by compulsion often met with failure. With the Japanese invasion of New Guinea in 1941, there was increased awareness by the Australian Government of the importance of Papua and New Guinea as a colony and a buffer-state. As the war drew to a close a 'new deal' for the people of Papua and New Guinea was proposed. War damage was to be repaired, and the colony prepared for a future independence. With this came agricultural policies aimed at improving the nutrition and living standards of the native people by encouraging mixed farming on individual smallholdings. These would be capable of producing adequate subsistence, in addition to sufficient cash crops to obtain the money necessary to satisfy other desires and to pay taxes (Dick and McKillop 1976). In the field of health, a programme of hospital construction was initiated in 1947, Port Moresby General Hospital being completed in 1957 (Gunther 1972).

Although gold mining was pursued from the outset in colonial New Guinea, it was not until after World War I that production became economically significant. With the discovery and development of the substantial Bulolo and Edie Creek goldfields of Morobe Province came other developments, notably that of air transport. This was instrumental in gaining access to the goldfields and allowing the import of technological infrastructure. Subsequently, air transport was instrumental in the opening up of the New Guinea Highlands to European contact and economic development (Souter 1961).

Between the First and Second World Wars mining development was encouraged by the administrations of both Territories of Papua, and New Guinea. Early in this period, gold mining was the most important source of income to New Guinea, although the Papuan goldfields declined in importance with only one rich goldfield, Misima, still operating at the outbreak of World War II. Mining ceased during the war and was not resumed in most places after 1945, since the cost of rehabilitation and restoration of mining and plant operations was largely deemed uneconomical. However, in 1969 there were still two companies operating at profit, while in 1967–8, indigenous miners with small-scale operations in East and West Sepik, Eastern and Western Highlands and Morobe contributed about a quarter of the total 28 000 oz of gold exported in this period (Tudor 1969).

With a new conservative coalition government in Australia in 1949, there was increased emphasis on economic development in New Guinea. Although the rehabilitation of the goldfields was deemed unprofitable, the government was again much more sympathetic to expatriate plantations, and to village-based cash cropping. The first major post-war promotion of plantation crops was on the Gazelle Peninsula of New Britain in 1951, where village growers were encouraged to plant cocoa and to enlarge their coconut groves. Furthermore, promotion of coffee began in the highlands in 1953, and between 1952 and 1954 the Australian government opened the highlands to Australian settlers. A 'land-rush' took place and the settlers planted sufficient alienated land to form the basis of a viable coffee industry (Finney 1972). A village level coffee extension programme quickly followed this plantation expansion and expanded the export of primary products, almost eight-fold, between 1948 and 1966 (Table 5.5). Coconuts and coconut products remained the most important cash crop, forming about 40% of total export earnings in both 1948 and 1966, while coffee became the second most important cash crop, forming 18% of export earnings in 1966, from being totally absent from the economy in 1948. Gold declined in importance, from supplying 27% of all export earnings in 1948, to only 2% in 1966, this being a reduction by half in total earnings from gold, in an economy that had expanded enormously in copra, rubber, cocoa, coffee and timber production.

Concomitant with the onset of colonial administration was industrialisation (Wilson 1972) and urbanisation (Harding 1981). Manufacturing industries were set up originally to process primary products, to maintain and repair vehicles, and to provide a few consumer services. Of the few facilities that were established before World War II, most were in Rabaul, New Britain, or on the

TABLE 5.5. Value of exports (US$ million) by main commodity classes, 1948 to 1975 (from Ryan 1972; National Census 1980)

	1948	1951	1954	1957	1960	1963	1966	1975
Coconuts and coconut products	2.7	10.4	16.7	14.2	20.6	14.8	20.9	62.1
Rubber	0.3	1.6	1.2	2.3	3.0	2.3	2.6	3.6
Cocoa	0.03	0.2	0.5	0.9	3.3	5.9	4.4	56.1
Tea	na	na	na	na	na	na	na	5.4
Coffee	0	0.02	0.1	0.4	1.4	4.0	8.8	47.0
Palm oil	na	na	na	na	na	na	na	9.5
Pyrethrum	0	0	0	0	0	0	0	0.3
Fish	na	na	na	na	na	na	na	14.1
Timber	0.01	0.05	0.4	2.4	3.3	2.7	3.7	13.6
Gold	1.7	2.7	2.8	2.5	1.3	1.3	0.9	4.8
Copper	0	0	0	0	0	0	0	331.3
Total	6.3	16.7	24.1	26.2	37.6	36.3	49.8	547.8

na : not available

TABLE 5.6. Value of imports (US$ million) by commodity classes between 1954 and 1966 (from Ryan 1972)

	1954	1957	1960	1963	1966
Food	8.0	9.8	10.7	13.5	22.6
Clothing, footwear, textiles	2.5	3.2	4.0	4.9	8.6
Machinery, metals, metal manufactures	6.0	9.5	9.5	12.8	31.6
Transport equipment, mineral fuels and lubricants	3.7	4.9	5.3	7.9	16.4
Other manufactured goods, drugs and chemicals	4.9	8.2	8.5	12.0	21.4
Total	27.6	39.2	41.6	56.5	109.1

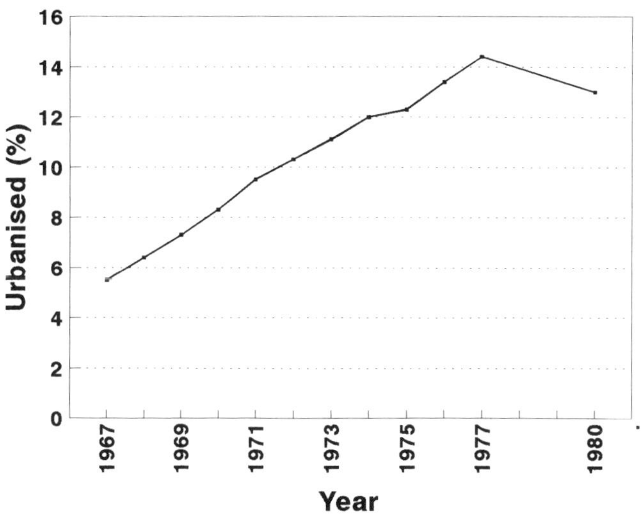

Fig. 5.7. The proportion of the total population of Papua and New Guinea urbanised, 1967–80. (From National Census 1981)

goldfields. Between 1960 and 1966, the number of manufacturing or processing units doubled, from 209 to 407. This still represented a very low level of industrialisation, since the 10 000 workers engaged in these production units in 1966 was less than 10% of the total work force of the country (Wilson 1972), the rest being employed by the public service. Administrative centres grew in size and the consumption of imported foodstuffs became increasingly important (Spencer and Heywood 1983). The value of food imported between 1954 and

1966 almost trebled, from US $ 8 million to $ 23 million, but this was a reduction in the proportion of overall value of imports, from 29% to 20%, as the value of all other imports increased more than four-fold (Table 5.6).

Although public servants were initially European, there was a steady increase in the number of indigenous public servants, this forming the major source of wage employment for the indigenous population. This section of the population was alienated from the land, just as indentured plantation workers had been in the past, and were extremely susceptible to modernization. Urban markets, selling traditional foods, were a product of the colonial administration. Village people were encouraged to 'market garden' and sell their surpluses at urban markets (Dick and McKillop 1976). However, prices of these foods were high, urban wages were low, and the best value for money could be obtained by buying imported rice (Ulijaszek 1993).

The urban population, although growing (Fig. 5.7), was still small compared to most other developing countries. The urban population plateaued between 13–14% of the total population between 1976 and 1980 (Policy Planning and Research 1991). By comparison, the average level of urbanism in all less developed countries was 29% in 1980 (Department of International Economic and Social Affairs 1986).

In 1968, the first five year plan for the development of Papua New Guinea was announced. This built on the report of the World Bank Mission of 1963, in which it was concluded that while substantial economic growth might be possible over the following 5 to 10 years, real economic viability would take several decades. It also stressed the need for continued and increasing participation of European economic interest in the Territory. The purpose of the 1968 report was to quicken the progress of the territory toward economic self reliance, by expansion of production in all areas. This included increasing the planting of tree crops by the indigenous population by 50%, increasing five-fold the number of beef cattle raised, to expand timber production and exports three-fold, and to double the manufacturing output. In addition, it recommended substantial increases in spending on roads, bridges and other transport facilities. It is not clear whether these aims had a broader humanitarian aim, other than the implicit belief of economists that increased income per capita and possession of resources could be directly equated with human well-being.

Post-independence economic development

In 1975 the Australian Territory of Papua and New Guinea became independent Papua New Guinea. The pre-independence plans sought to reduce reliance on external support for its social, political and bureaucratic structures. Although cash cropping had brought increasing export revenues to the country (Table 5.5), these were far from adequate to reduce dependency on overseas aid. Although gold mining had gone into decline, the possible exploitation of other mineral reserves had become apparent prior to independence, with the prospecting for copper in Bougainville in 1964. A lease for copper mining was granted in 1969, with production starting in 1972. Thus, before independence the mining of copper was already underway at Bougainville, and by 1975 was the dominant national commodity, supplying 60% of total export earnings.

Prospecting for minerals had also been underway in the Star Mountains of the Western Province from the late 1960s onward. After the discovery of gold and copper deposits, a consortium of internationally based companies formed the Ok Tedi Mining Company in 1981, which immediately began the development of mining infrastructure, and started mining deposits of gold in 1984. Further mineral exploitation of gold and tin reserves have since been resumed at Misima and undertaken at Polgera, Lihir and elsewhere, respectively.

THE IMPACT OF ECONOMIC DEVELOPMENT AND MODERNISATION ON HEALTH IN PAPUA NEW GUINEA

Modernisation in New Guinea is of comparatively recent origin, and is not as advanced as in other less developed countries. As elsewhere, implicit in the New Guinea development planners' rationale is the belief that, economic development and modernisation will lead to improvements in well-being of most of the population. When addressed at all, health is viewed in the context of improved well-being, as assessed by increased per capita GNP. This view can be questioned by examining different aspects of health and well-being in populations at different stages of modernisation. In New Guinea there are populations that have only recently been exposed to the western technological culture complex, communities who have practised cash cropping for less than two generations, and groups that are clearly urbanized.

Early health effects of culture contact

There are no health statistics on the indigenous populations of the New Guinea island at the time of European contact and soon after, so the health effects of contact cannot be stated with any clarity. However, epidemics of infectious diseases often occur in populations which have had no previous exposure to them, and to which they have little resistance (Swedlund and Armelagos 1990). Changed patterns of post-contact infectious disease have been described for populations in South America (Wirsing 1985), Mexico (Vargas 1990), and Highlands New Guinea (Dennett and Connell 1988). Although prehistoric populations in transition from one pattern of subsistence to another have been shown to have changed types and frequencies of skeletal pathologies (Armelagos 1990), it is impossible to determine whether the arrival of new diseases manifested themselves as epidemics. However, virgin soil epidemics of measles have been described for populations in the Amazon valley in 1749, among the Indians of the Hudson Bay Territory in 1846, the Hottentots of the Cape of Good Hope in 1852, the Tasmanians in 1854 and 1861, the Mauritanians in 1874, the Fiji Islanders in 1875, and the Rotuma Islanders in 1911 (Brincker 1938; Glanville Corney 1913; Babbot and Gordan 1954), while Darwin (1845) noted in an 1836 entry in his journal of the voyage of the HMS *Beagle* on the gradual extinction of the Australian Aboriginal people due to the introduction of measles by Europeans. In New Guinea, evidence of virgin soil epidemics comes from administrative and medical reports.

Annual Reports of British New Guinea, and Annual Reports to the League of Nations, give brief details of a smallpox epidemic in Central Papua in 1894, and a dysentery epidemic in the New Guinea goldfields with above 50% case fatality in 1921 (Vines 1972). Dysentery flourished in the mining camps and plantations, while tuberculosis was introduced to New Guinea by expatriots, with Polynesian teachers and pastors being among the most thoroughly infected (Denoon 1989). The link between colonialism and tuberculosis was made in the 1930s, when it was observed that the disease was virtually absent from the highlands, which were only recently explored by Europeans, but common along coastal regions and most common in Hanuabada, a series of urban villages within the boundaries of Port Moresby, the oldest European settlement and the capital of the Territory (Clements 1936). This pattern of tuberculosis prevalence was largely consistent with the history of contact with outside populations in Papua and New Guinea. Furthermore, serological surveys of measles antibodies in various populations in East and West New Guinea between 1959 and 1960 showed that only the remotest groups in Western New Guinea had not previously been exposed to this disease (Adels and Gajdusek 1963), again following a colonial pattern of disease introduction.

The most complete account of the health effects of culture contact in New Guinea comes from the Star Mountains of the Western Province, where the Wopkaimin population became integrated in the modernisation process very recently, as result of the development of the Ok Tedi gold and copper mining operation (Ulijaszek and Pumuye 1985; Lourie *et al.* 1986; Lourie 1987; Jones *et al.* 1987; Ulijaszek *et al.* 1987, 1989; Schuurkamp *et al.* 1987*a*, *b*; Taufa *et al.* 1990; Hyndman 1990; Schuurkamp 1992).

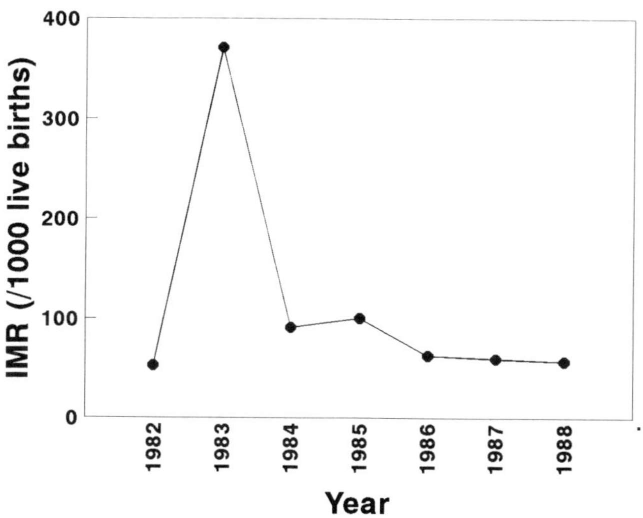

Fig. 5.8. Infant mortality rates among the Wopkaimin, Papua New Guinea, 1982–1988. (From Taufa *et al.* 1990)

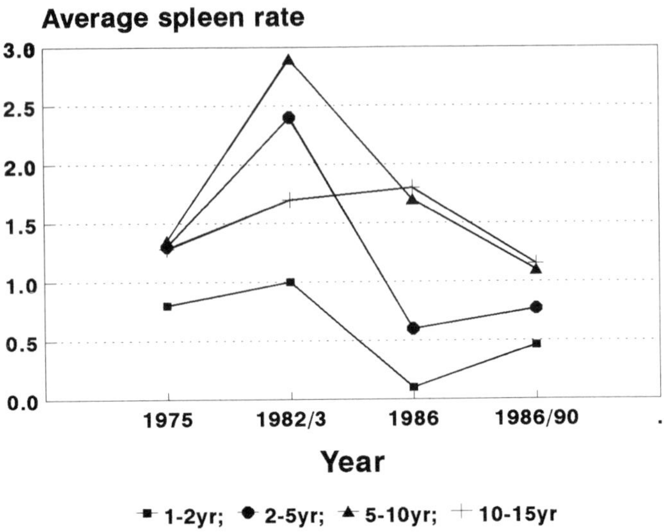

Fig. 5.9. Average spleen rates of Wopkaimin children, 1975–1990. (Data from Papua New Guinea Institute of Medical Research Star Mountains Survey 1975 (unpublished); Lourie *et al.* 1986; Schuurkamp 1992)

Evidence for post-contact infection-related mortality among the Wopkaimin comes from reports of: 1) dysentery, which in 1968 claimed 5–6% of the population in two census parishes (Eggleton

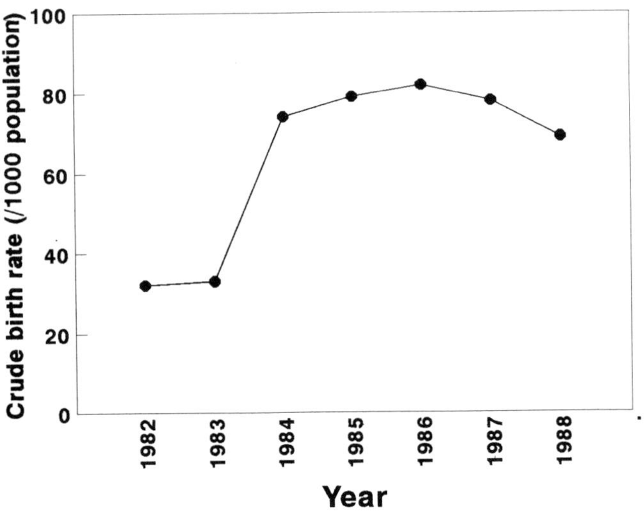

Fig. 5.10. Crude birth rate, Wopkaimin, 1982–1988. (From Taufa *et al.* 1990)

1968); and 2) many deaths in 1971 as a result of what was described as whooping cough (McGrath 1972). Furthermore soon after the onset of the mining operation at Ok Tedi in 1981 there was increased infant mortality due largely to a measles epidemic (Fig. 5.8) (Taufa *et al.* 1990). This may have been a virgin soil epidemic; however, since measles antibodies were found in a related Star Mountains population in West New Guinea in 1959 (Adels and Gajdusek 1963) it may not have been. At that time, this was an isolated population that had first contact with Europeans only a decade previously, and had only recently built an airstrip. In 1981, the Wopkaimin had a longer history since European contact, which first took place in 1957. The possibility that the Wopkaimin had been exposed to measles through contact with neighbouring populations prior to the onset of modernisation therefore cannot be discounted.

There was also increased exposure to malaria, probably as a result of road contact with the more lowland malarious region around Kiunga and Ningerum. Figure 5.9 shows average spleen rates in children by age group between 1975 and 1990. Spleen rates were greater in 1982/3 than in 1975 for all age groups, with substantial declines taking place in all age groups except 10–15 years between 1982/3 and 1986. By 1986/90, spleen rates were slightly lower than they had been in 1975. Thus prior to the onset of the mining operation, the Wopkaimin lived in a stable malarious area. After this time, malaria was imported through workers recruited elsewhere in PNG (Schuurkamp *et al.* 1987*a*). The reduction in spleen rates after 1982/3 reflects a reduced malarial load in the population due to malaria control activities implemented by Ok Tedi Mining Limited (Schuurkamp *et al.* 1987*a*, *b*; Schuurkamp 1992).

There was also an increase in birth rate after 1983 (Fig. 5.10) (Taufa *et al.* 1990), a demographic event which runs counter to fertility transition models for developing countries (Caldwell 1982). Taufa *et al.* (1990) suggest that a change in domestic arrangements may have accounted for this increase. In the old villages, men slept separately from the women and children, in the men's house; in the course of modernisation, men and their wives adopted the practice of sleeping together in their own house. Two supplementary reasons are offered here: 1) with the influx of workers from other regions of PNG, a 'novel male' effect might have operated, and some of the increased fertility may have arisen from female promiscuity as a result of exposure to these outsiders; and 2) derepression of impaired ovulatory function as a result of improved nutrition and reduced workloads may have made a greater proportion of coital episodes fertile.

Dietary and nutritional change was pronounced. Wopkaimin were traditionally shifting hunter-horticulturalists, with a traditional staple food of taro (*Colocasia* species). Recent times have seen a shift to sweet potato (*Ipomea* species) and Xanthosoma taro cultivation among Wopkaimin living away from the Ok Tedi mine site, with a considerable intake of western storefoods, notably rice, among those living closer to the mine and working, or having relatives working for cash income. In 1984, Wopkaimin mine workers obtained 64% of their dietary energy from such sources, while their wives obtained 38% from the same sources (Ulijaszek *et al.* 1987). This has had a significant impact on nutritional status, with adults being heavier,

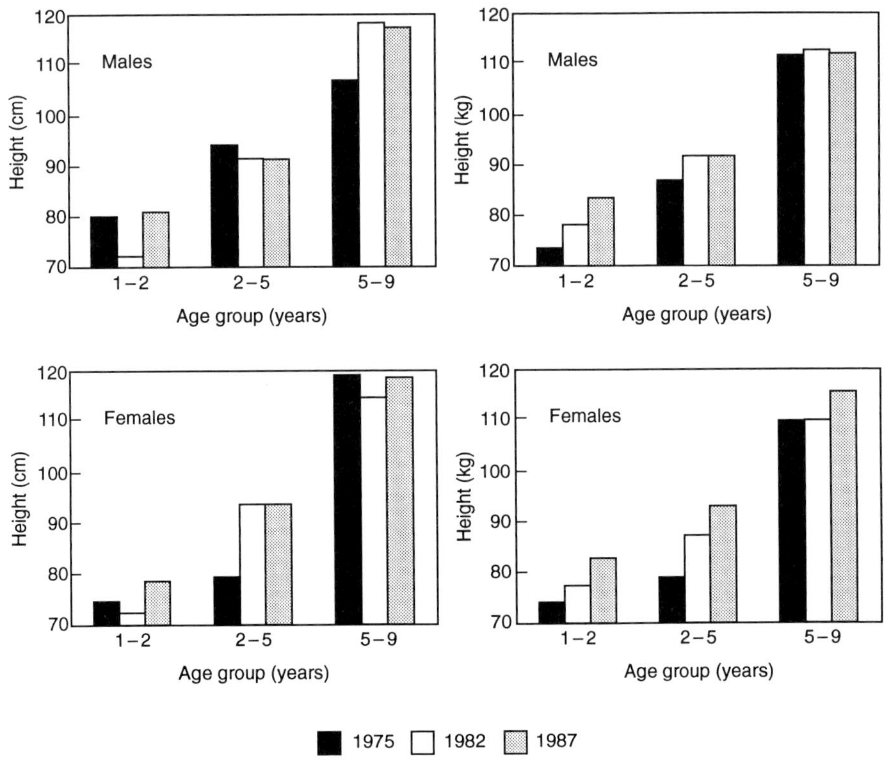

Fig. 5.11. Mean heights and weights of Wopkaimin children by age group and sex, 1975–1987. (Data from Papua New Guinea Institute of Medical Research Star Mountains Survey 1975 (unpublished); Lourie *et al.* 1986; unpublished Ok Tedi Health and Nutrition Project 1987)

and children being heavier, and in some sex and age groups taller, than prior to mining (Fig. 5.11).

Thus the short term effects on health of culture contact for the Wopkaimin have been: 1) a disrupted disease ecology, with exposure to new pathogenic agents, and increased exposure to ones which were previously limited; 2) dietary change, even for Wopkaimin not directly involved in the cash economy; 3) nutritional improvements for those Wopkaimin and their families directly involved in the cash economy, mostly as wage earners; and 4) increased birth rates, possibly as a result of increased fecundity.

TABLE 5.7. Dietary change in Simbu Province, Papua New Guinea: energy and protein intakes (from Harvey and Heywood 1983*a*)

	Energy				Protein	
	1956		1981		1956	1981
	(MJ)	(kcal)	(MJ)	(kcal)	(g)	(g)
Adult males	10.4	2485	9.8	2340	37.2	43.3
Adult females[a]	7.0	1684	7.8	1864	30.1	35.8
			8.0	1903		37.9
Children:						
15–19 yr	7.3	1735	11.3	2710	25.9	50.9
10–14 yr	5.6	1340	7.9	1894	22.4	38.8
5– 9 yr	5.1	1220	6.6	1580	24.3	32.6
3– 4 yr	3.8	920	6.2	1473	16.6	30.0
1– 2 yr	3.1	730	3.4	810	10.8	21.6

[a] 1981 measurements taken at two different times of year

TABLE 5.8. Cash cropping in Simbu Province, Papua New Guinea. Proportion of dietary energy from different foods in 1956 and 1981. (from Harvey and Heywood 1983*a*)

Food or food group	1956 (%)	1981 (%)
Sweet potato	76	53
Other starchy tubers	7	1
Cereals and grains	4	22
Fish and meat	4	6
Leafy vegetables	1	2
Other vegetables and fruit	8	10
Fats and oils	0	2
Alcohol	0	2
All store foods	0	26

The health impact of longer-term development and modernisation

The health effects of development in populations which began modernisation earlier are best documented for rural groups practising cash cropping. In particular, the effects of cash cropping on nutritional status in the Simbu Province of the highlands have been documented (Harvey and Heywood 1983*a*, *b*), and are examined here.

Sweet potato became the dominant staple of the New Guinea Highlands about 400 years ago (Ulijaszek 1994). However, the instigation of smallholder coffee cash cropping in the 1950s after European culture contact and establishment of colonial administration in the highlands brought changes in food use. In Sinasina District, Simbu Province, the number of coffee trees under indigenous cultivation increased more than five-fold between 1956 and 1962, from about 50 000 trees in 1956, and increasing by a third again between 1962 and 1968 (Hatanaka 1972). Cash cropping allowed new opportunities for resource acquisition, by selling commodities and buying imported goods, the most important of which was food (Heywood and Jenkins 1992).

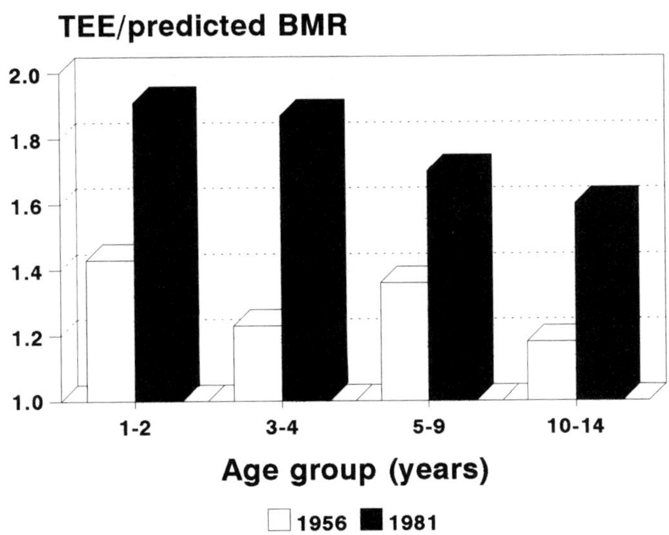

Fig. 5.12. Dietary change in Simbu Province, 1956–81. Energy intakes expressed as multiples of basal metabolic rate (predicted from body weight, using the equations of Schofield 1985), by age group. (Data from Harvey and Heywood 1983*a*)

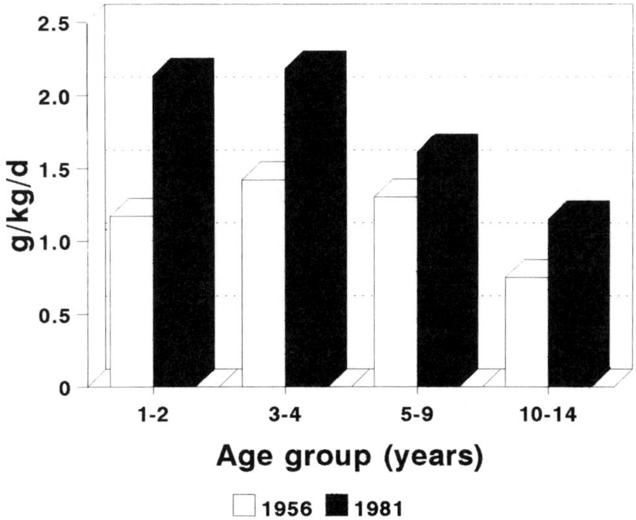

Fig. 5.13. Dietary change in Simbu Province, 1956–81. Protein intakes per kg of body weight, by age group. (Data from Harvey and Heywood 1983*a*)

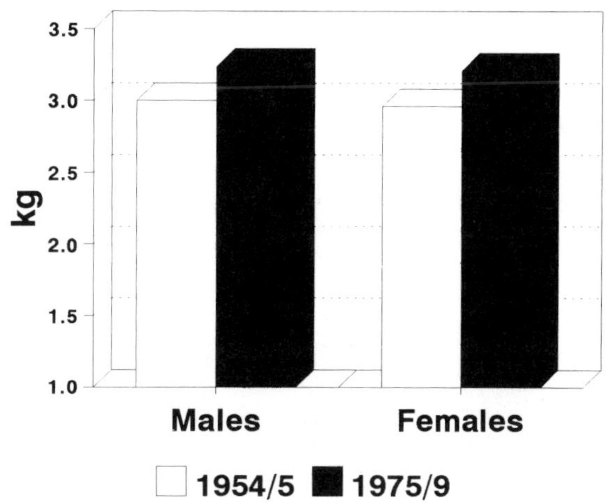

Fig. 5.14. Mean birth weight of Simbu infants, 1956 and 1981. (From Venkatachalam 1962; Harvey and Heywood 1983*b*)

Table 5.7 gives mean energy and protein intakes of adults and children in 1956 and 1981, and Table 5.8 gives the proportion of dietary energy obtained from different foods for one Simbu

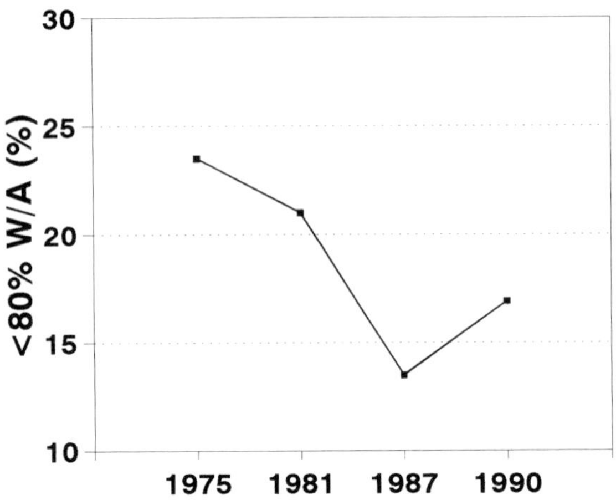

Fig. 5.15. Proportion of children aged 0–5 years below 80% of weight for age (National Center for Health Statistics 1977) Simbu Province, 1975–1990. (From Harvey and Heywood 1983*b*; Policy Planning and Research 1991)

population. As a result of coffee cash cropping prosperity, there was an increase in the consumption of cereals and grains, predominantly of imported rice. In 1956, cereals and grains provided only 4% of the total energy intake, the grains being predominantly maize and indigenous grain legumes. In 1956, sweet potato and other starchy tubers supplied 83% of the dietary energy. In 1981 this had dropped to 54%, the difference being made up almost totally by imported storefoods. Traditional sources of fish and meat had been largely

TABLE 5.9. Mean heights and weights of adults in Simbu Province

Year	N	Height Mean		(SD)	Weight Mean		(SD)
Females (non-pregnant)							
1956	203	150.1		(6.4)	47.9		(5.1)
1981	839	151.4		(5.0)	51.5		(5.7)
Student's t-test			2.70*			8.80**	
Males							
1956	80	159.9		(5.4)	55.1		(5.6)
1981	17				61.1		(5.7)
Student's t-test						3.95**	
		*< 0.01		**p < 0.001			

Sources: Venkatachalam 1962; Harvey and Heywood 1983*a,b*

replaced by imported tinned meat and fish. Other new items in the diet were fats and oils, and alcohol, predominantly from beer consumption.

This dietary change led to increased intakes of both energy and protein in women and in children of all ages, and to increased intakes of protein in adult males. Increases in energy intake, when expressed as a multiple of basal metabolic rate predicted from body weight (Schofield 1985) were greatest in the youngest age groups, 1–2, and 3–4 years (Fig. 5.12). Children below the age of 5 years were similarly favoured with respect to protein intakes (Fig. 5.13). There was also an increase in mean birthweight (Fig. 5.14) and in the mean height of women, and mean weights of adults of both sexes (Table 5.9) across this period. Although the proportion of children below 80% W/A declined between 1975 and 1987, it showed an increase between 1987 and 1990 (Fig. 5.15).

Infant and young child mortality rates had declined from 149 and 77 deaths per 1000 live births in 1971, to 87 and 47 deaths per thousand live births in 1980, respectively (Policy Planning and Evaluation 1991). In the absence of early demographic data it is impossible to make any statement about changing fertility rates across the period of cash crop modernisation in Simbu. However, the birth rate in 1980 was 26.1, the lowest in the country, while birth rates in all the Highlands provinces which had undergone similar changes were below the national average (Policy Planning and Research 1991). Thus it appears that populations further into the modernisation process experience some improvements in health, as indicated by nutritional and demographic markers, with possible decline in fertility.

The health impact of urbanism

There has been no concerted study of urbanisation and health in New Guinea. However, health problems in the urban centres are probably largely socio-economic in origin. Poor wages, high food prices and a shortage of land on which to make household gardens may predispose certain sectors of the urban population to nutritional deprivation. Demand for traditional staple foods in the towns has risen, as has price, even after allowing for inflation (Table 5.10). Imported rice has been displacing other staple foods in the urban areas because its price has remained more or less constant. However, urbanism may be also associated with economic opportunity, and greater access to free health care provision.

TABLE 5.10. Yearly growth in the real price of staple foods at urban markets in Papua New Guinea, 1971–1981. Figures are percentages above a baseline of 100 (from Spencer and Heywood 1983)

	Sweet potato	Taro	Plantain	Cassava
Port Moresby	+ 5*	+ 7*	+ 5*	nc
Lae	+ 6*	+ 12*	nc	nc
Rabaul	+ 6*	+ 8*	+ 5*	nc
Madang	nc	nc	– 3+	nc
Goroka	+ 6+	+ 14*	+ 6*	+ 4+

nc: no significant change
+p <0.05
* p <0.01

Studies of urban health are limited to health reports, and the occasional more detailed survey. A comparison of children's weights in Lae Town in 1958–72 (Malcolm 1970) and in 1982 (Kow 1987) shows the latter to be heavier, indicating an improvement in child nutritional status in one urban setting. In addition urban Bundi children have larger body size and begin puberty earlier than do rural Bundi (Zemel *et al.* 1993). Further evidence of the health effects of urbanism in PNG can be evinced from census data (Policy Planning and Evaluation 1991). In 1990, the proportion of children below five years of age in the National Capital District (NCD) below 80% weight for age was lower than the all PNG average, and lower than for Central Province in general (Table 5.11). Indeed, the figures for Central Province, which is the predominantly rural area surrounding the NCD, is higher than the national average.

TABLE 5.11. Proportion of children below 80% of weight for age, National Capital District, Central Province, and all Papua New Guinea, 1990 (from Policy Planning and Research 1991)

	Children below 80% weight for age[a]			
	0–1 yr		1–4 yr	
	Total n[b]	%	Total n[b]	%
National Capital District	41972	12.6	34091	17.7
Central Province	17289	31.1	27125	38.6
All Papua New Guinea	554347	16.3	1204922	27.9

[a] below 80% of the National Center for Health Statistics (1977) fiftieth centile
[b] number of clinic attendances, January–December, 1990

Table 5.12 gives a comparison of infant mortality rates for the NCD, all PNG, and the mostly rural Central Province. Again, values are much lower in the predominantly urban national capital. Table 5.13. gives vital rates for all PNG, divided by urban and rural residence. Birth rates are similar in both, but the urban death rate is half that of the rural population, largely as a result of the lower infant mortality rate, presumably because of better nutritional and health care provision in urban centres. As a consequence of the greater disparity between birth and death rates, the urban population growth rate is much larger than that of the rural population. At 2.9% per annum, this suggests a doubling time for the urban population of 25 years on this basis alone, excluding possible urban population increase due to rural–urban migration.

It would appear that on average, urbanism is better for health than is rural life. There is little information on the extent of social disparity

TABLE 5.12. Infant mortality rates, National Capital District, Central Province, and all Papua New Guinea, 1980 (from Policy Planning and Research Division 1986; Policy Planning and Research 1991)

	% of total population urban	Deaths per 1000 live births		
		Males	Females	Total
National Capital District	100	37	33	35
Central Province	4	61	55	59
All Papua New Guinea	13	78	66	72

TABLE 5.13. Vital rates, urban and rural Papua New Guinea, 1980 (from Policy Planning and Research 1991).

	Birth rate (a)	Death rate (a)	Population growth rate (% per annum)
All Papua New Guinea:			
Urban	36.7	7.7	2.9
Rural	33.9	14.2	2.0
Combined	34.2	13.2	2.1
Selected provinces:			
National Capital District	37.7	6.9	3.1
Central Province	34.0	13.4	2.1
Simbu Province	26.1	15.9	1.0

[a] per annum

in PNG, and none on the effects of modernisation on possible amplification of social and economic disparity. Given the rapid pace at which the country is modernising, biosocial research in this area is a high priority.

Relationships between economic development, modernisation, and intervention

The introduction of a monied economy and involvement in the global fiscal economy has set New Guinea on a path of interdependency and change. On the basis of studies in two regions of PNG (Harvey and Heywood 1983*a*,*b*; Lourie 1987; Ulijaszek *et al.* 1987; Schuurkamp 1992) it appears that there have been improvements in the health and nutritional status of populations undergoing modernisation, although there seems to be some reversal of this trend in Simbu. However, it is not clear to what extent health improvements have been due to economic development and the process of modernisation which has accompanied it, and to what extent they could be attributed to specific health interventions such as primary health care and immunisation. Furthermore, it is not clear whether the results of these studies can be generalised to the larger PNG population. To examine these issues, health indicators, standardised for projected

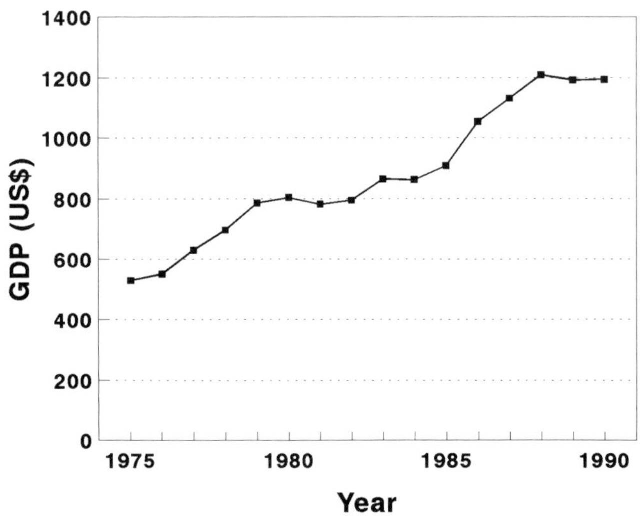

Fig. 5.16. Per capita gross domestic product (US$), Papua New Guinea, 1975–1990. (Calculated from data in: Policy Planning and Research Division 1986; Policy Planning and Research 1991)

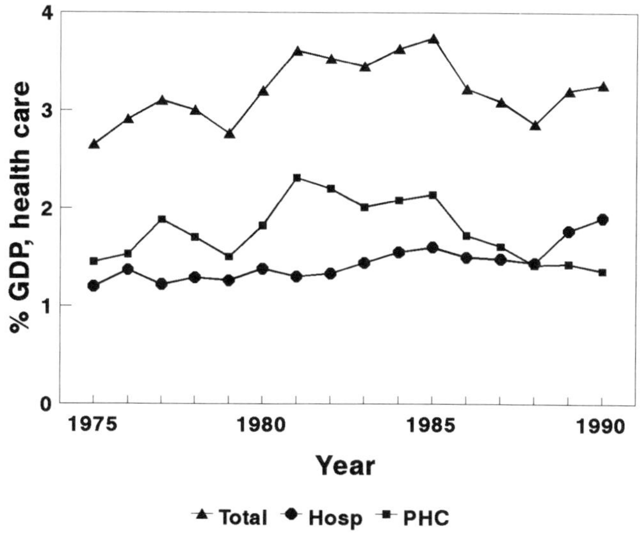

Fig. 5.17. Proportion of per capita gross domestic product (US$) spent in provision of health services, Papua New Guinea, 1975–1990. Total expenditure, expenditure on hospitals (Hosp), and on primary health care (PHC). (Calculated from data in: Policy Planning and Research Division 1986; Policy Planning and Research 1991)

change in population size from National Census (1981) data, were calculated from data published in health statistics reports for PNG (Policy Planning and Research Division 1986; Policy Planning and Research 1991) across time. The health indicators used included disease prevalence rates, immunisation coverage, and nutritional status.

Gross domestic product has increased more than three-fold between 1975 and 1990, outstripping population increase. However, total health expenditure has increased four-fold across this period. Per capita GDP (Fig. 5.16) has more than doubled, while the proportion of per capita GDP spent in health care provision has fluctuated across this time, but has generally increased (Fig. 5.17). Despite claims that primary health care is the way forward in controlling the problems of infectious disease and undernutrition, the increase in the proportion of per capita GDP spent in health care can be totally accounted for by increased spending on hospital facilities and care, since the trend line for hospital spending in Fig. 5.17 runs parallel to that of total spending, while that for primary health care spending is flat, between 1975 and 1990.

Despite the lack of increased spending in primary health care as a proportion of GDP, overall spending in primary health care has at

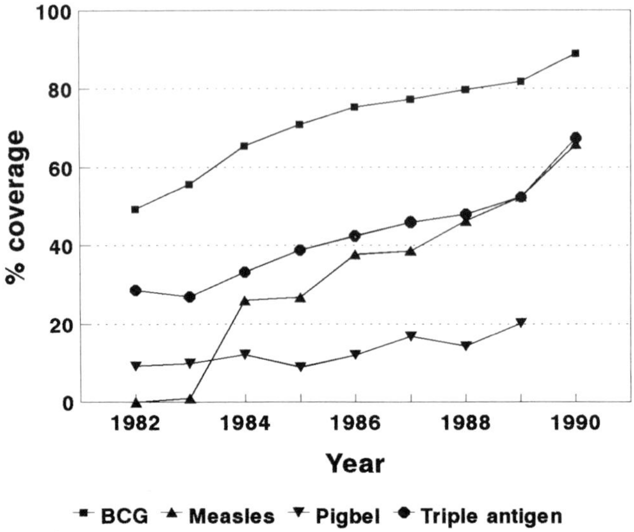

Fig. 5.18. Immunisation coverage rate (per cent), 1982–1990. Values for pigbel and triple antigen are for the proportion of children receiving a complete course (three doses) of vaccine. Coverage with pigbel vaccination is low, since the vast majority (more than 90%) of immunisations are focused in five provinces: Southern, Western, and Eastern Highlands, Simbu and Enga Provinces. (From Policy Planning and Research Division 1986; Policy Planning and Research 1991)

least been maintained as a proportion of overall increased GDP, and improved performance in this sector can be judged by the proportion of the population covered by the immunisation programme to children below the age of one year (Fig. 5.18). BCG coverage increased from 49% in 1982, to 89% in 1990, while triple antigen coverage increased from 29% to 67% across the same period. Measles vaccination, which was non-existent in 1982, reached 66% in 1990. Pigbel vaccination increased from 9% of the PNG population in 1982, to 20% coverage in 1989. However, this value gives a distortedly low estimate of coverage in the parts of PNG with the highest incidence of pigbel. In the five Highlands provinces, Southern, Western, Eastern Highlands, and Simbu and Enga Provinces, the proportion of children below 1 year of age having received all three doses of vaccine was 26% in 1982, 38% in 1986, and 67% in 1989 (Policy Planning and Research Division 1986; Policy Planning and Research1991).

The success of the immunisation programme is reflected in the reduced incidence rates of measles and pigbel (Table 5.15), and to

TABLE 5.14. Infection rates for measles, National Capital District, and Central Province, 1984–1990. Rates are based on reported cases of measles and projections of population size from the 1980 National Population Census (from Policy Planning and Research Division 1986; Policy Planning and Research 1991)

	Rate per 100 000 population	
Year	National Capital District	Central Province
1984	1182	294
1985	330	66
1986	1702	107
1987	723	265
1988	183	51
1989	1246	155
1990	119	117

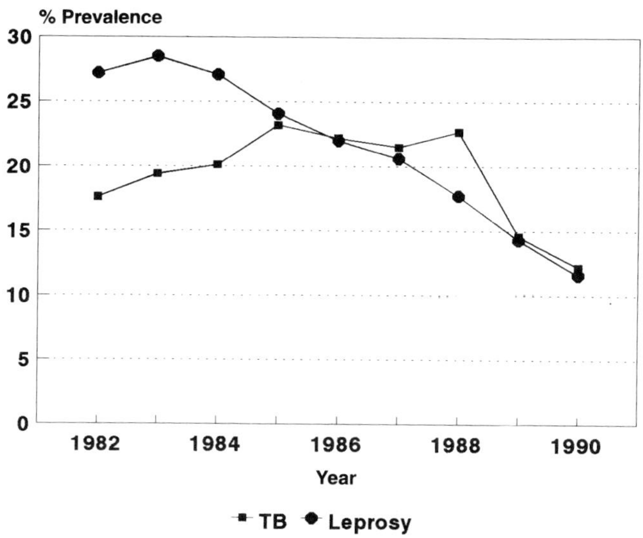

Fig. 5.19. Prevalence of tuberculosis and leprosy per 10 000 population, 1982–1990. (From Policy Planning and Research Division 1986; Policy Planning and Research 1991)

some extent the relatively recent decline in tuberculosis prevalence (Fig. 5.19). Although fluctuating from year to year, the trend in measles infection is downward, with a three-fold decrease in incidence between 1982 and 1990. Pigbel has also seen a three-fold decline in incidence. Tuberculosis and leprosy control is carried out by the detection and follow-up of individuals with these diseases by

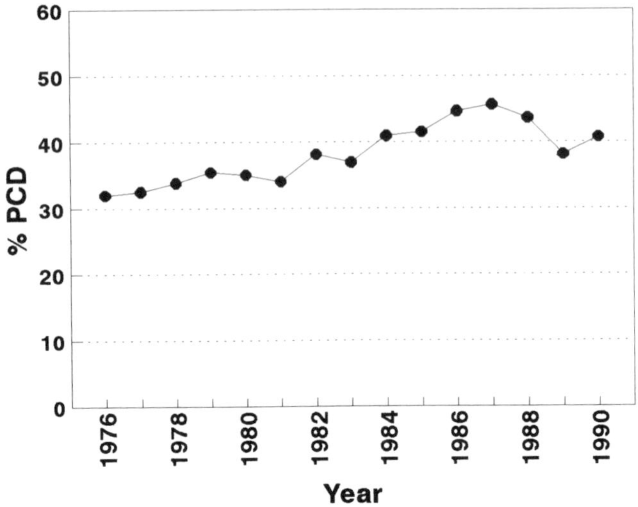

Fig. 5.20. Positivity rate of passive case detection of malaria, 1976–1990, percentage of slides positive. (From Policy Planning and Research Division 1986; Policy Planning and Research 1991)

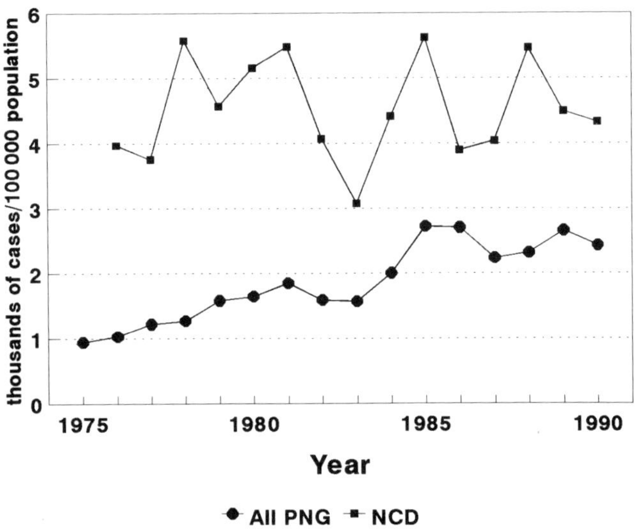

Fig. 5.21. Rates of diarrhoeal infection (thousands of cases per 100 000 population) in all Papua New Guinea (PNG) and the National Capital District (NCD), 1975–1990. (From Policy Planning and Research Division 1986; Policy Planning and Research, 1991)

workers employed in the primary health care sector, and so the declining prevalence of both these diseases reflects the success of these programmes, as does the reduction in infant mortality rate, from 134 per thousand live births in 1971, to 72 in 1980. However, two classes of disease, diarrhoea and influenza, both of great public health significance but which have not been the focus of primary health care action, have increased in incidence between 1982 and 1990 (Table 5.16). Furthermore, the prevalence of malaria, as identified from passive case detection, has steadily increased between 1976 and 1990 (Fig. 5.20).

The incidence rate of diarrhoea has increased more than two-fold between 1975 and 1990, from 947 cases per 100 000 population in 1975, to 2428 per 100 000 in 1990 (Fig. 5.21). Although these figures are for the total number of cases, in 1990, 1224 cases, or 50% of the total, were for children aged 1 to 4 years, and it seems reasonable to assume that similar proportions of total morbidity due to diarrhoea in young children and adults might prevail in other years. The incidence of diarrhoea in the NCD, a representative urban population, was much higher than in the population as a whole. This might be due to greater reporting in the NCD than elsewhere in the country, and/or could reflect a bias in the reporting of diarrhoeal infections toward older individuals.

However, if these figures reflect a true difference in young child diarrhoeal morbidity between urban and rural populations, there are some contradictory observations. Firstly, the incidence of diarrhoeal disease has not increased in the NCD between 1975 and 1990, as it did in the country as a whole. Furthermore, it was estimated in 1983 that the proportion of the rural population with safe water was only 10%, compared with 55% for the urban population (Policy Planning and Research Division 1986) which runs counter to expectation on the basis of diarrhoeal morbidity. In addition, the proportion of children undernourished was lower in the NCD than in the country as a whole (Table 5.12). On the basis of the known associations between diarrhoeal infection and undernutrition in young children (Tomkins 1983; Rowland *et al.* 1988; Briend *et al.* 1989), it might be expected that higher diarrhoeal incidence in the urban population would be associated with higher levels of undernutrition. However, results from several population-based studies around the world suggest that diarrhoea alone may not necessarily be associated with undernutrition, and that poorer environmental conditions may predispose a population to diarrhoea-nutritional status interactions (Tomkins 1986). It may not be diarrhoeal morbidity alone that predisposes to undernutrition, but the complex relationships between environmental quality, infection, and the management of that infection. At the population level, improved management of diarrhoea could result in lower case fatality rates, which might be manifested as a greater proportion of malnourished children. However, this cannot explain the urban–rural difference in diarrhoeal incidence and child nutritional status.

It has been suggested that the immune suppression of some 3–4 months duration which follows infection with measles virus may contribute to diarrhoeal morbidity (Tomkins 1986). This could explain the pattern observed in the NCD, since measles morbidity was generally considerably higher there than either in the country as a whole, or in the predominantly rural Central Province population (Tables 5.14 and 5.15), despite the greater immunisation coverage in the NCD. The paradoxical relationship between these patterns of infection and nutritional status might be a function of the treatment and management of both measles and diarrhoea, and the duration and severity of diarrhoeal infection in the urban context. It can only be assumed that increased survival of these diseases is not associated with impaired nutritional status at the population level, and/or that the nutritional status of NCD infants is better than the national average from birth. If this were the case, infectious disease experience

TABLE 5.15. Infectious disease rates, Papua New Guinea, 1982–90. Rates are based on reported cases of diseases and disease categories, and projections of population size from the 1980 National Census (from Policy Planning and Research Division 1986, Policy Planning and Research 1991)

	Infectious disease rate per 100 000 population			
Year	Diarrhoea	Influenza	Measles	Pigbel
1982	1597	1410	410	26
1983	1575	2162	314	15
1984	2008	1723	302	16
1985	2724	3043	169	9
1986	2705	1752	556	12
1987	2237	1693	293	8
1988	2318	2497	183	7
1989	2658	3749	268	7
1990	2428	2377	122	7

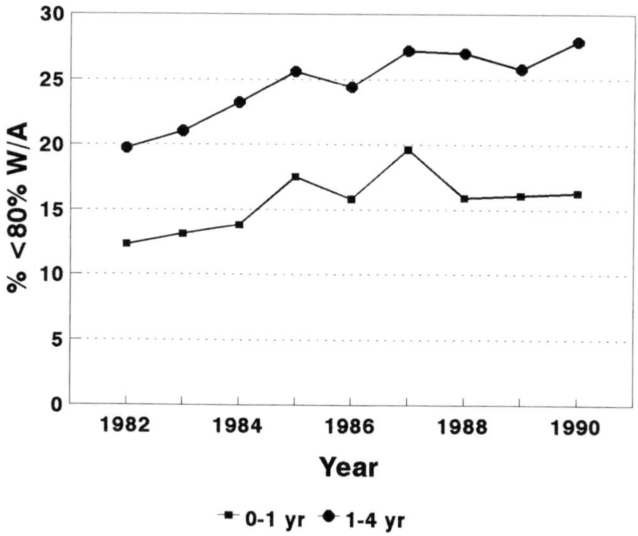

Fig. 5.22. Proportion of children considered undernourished on the basis of weight for age (W/A) below 80% of the National Center for Health Statistics (1977) fiftieth centile, 1982–1990. (Calculated from data in Policy Planning and Research Division 1986; Policy Planning and Research 1991)

would have less impact on child growth and nutritional status in the NCD than elsewhere.

The country-wide prevalence of undernutrition increased between 1982 and 1990 (Fig. 5.22), and may have been associated with the observed increase in diarrhoeal morbidity in the country as a whole, under one of two conditions. The first is if the availability of medical intervention and environmental quality were both low, resulting in an overall high level of diarrhoeal infection in the young child population. The second would be if improved management of diarrhoeal infection led to lower case fatality rates, but poorer nutritional status in children who survived diarrhoeal infection. The very low proportion (10%) of the rural population deemed to have access to safe water in 1983 is evidence of poor environmental quality. It might be that the combination of poor environment, improved management of diarrhoeal infection and lower case fatality rates may have resulted in a greater number of survivors, albeit with poorer nutritional status. It is also possible that the nutritional status of rural infants is poorer than that of urban infants from birth, and that diarrhoeal infection has greater impact on nutritional status, as a consequence. Furthermore, relationships between measles and diarrhoea may have been less important in the rural than the urban population, since measles rates between 1984 and 1990 were on average about three-fold lower in all-PNG, compared to the NCD (270/100 000 (all PNG), against 784/100 000 (NCD)). If so, the decline in measles morbidity outside of the NCD since 1982 is unlikely to have had much impact on nutritional status, since rates were low compared with rates of diarrhoeal infection even in 1982, and the proportion of children likely to have suffered the nutritional consequences of post-measles diarrhoea between 1982 and 1990 is likely to have been small.

Thus it would seem that the health and nutritional improvements associated with modernisation in the two closely studied populations described in this chapter are not representative of the country as a whole, where nutritional status has declined, and diarrhoeal incidence and malarial prevalence have increased, in the face of both increased GDP and greater spending on health care. However, immunisation coverage has increased, and the prevalence of tuberculosis and leprosy have both declined, as have the incidence rates of measles and pigbel.

The extent to which development, as measured by increased per capita GDP, has led to health improvement, and the extent to which this might be attributed to direct health interventions is difficult to determine, but is an important issue for public health planning. Although health sector spending in PNG has increased ahead of per capita GDP, most, if not all, of this increase has gone to support the provision and maintenance of hospital infrastructure and care, and not primary health care. The primary health care interventions with the highest profiles in PNG are immunisation, the follow-up and treatment of individuals with tuberculosis and leprosy, and malaria control. Increased immunisation coverage across the 1980s was probably only possible as a consequence of increased GDP, since the level of spending in primary health care has increased in tandem with per capita GDP; the declining prevalence and incidence of diseases for which immunisations are available suggests that this intervention has been effective. However, some diseases which cannot be immunised against are increasing in prevalence or incidence, including diarrhoea and malaria. Should effective vaccines be developed, it can be predicted that diarrhoea incidence and malaria prevalence would also decline.

Although the interactions between nutritional status, diarrhoeal disease and nutritional outcome are complex, it might be expected that increased material well-being would result in a decline in the incidence of diarrhoea and an improvement in nutritional status. This is not the case of PNG in general, despite case-studies suggesting that modernisation results in health improvement in both the Highlands (Harvey and Heywood 1983*a*, *b*; Dennett and Connell 1988) and the Highland fringe (Ulijaszek *et al.* 1987). Thus it would appear that economic development *per se* has little influence on health, except that it allows greater medical intervention. However, the two are closely tied, and interventions such as health education, which are largely behavioural and are difficult to evaluate, could have considerable influence on patterns of disease and nutritional status, both of which are also strongly influenced by economic well-being. In addition, the achievement motivation associated with economic development may also be associated with greater susceptibility to behavioural change in other spheres. For example, those individuals seeking economic improvement are also more likely to respond to health interventions, whether direct or indirect.

CONCLUSIONS

The issue of economic development and modernisation against intervention as policy for health improvement is complex; health intervention could take place independently of economic development. However, in the New Guinea context, development has made intervention possible. Significant levels of economic development were established only recently, and the health patterns and the extent and nature of health intervention observed in PNG to the present day are in large part the outcomes of the colonial past. Thus, health, development and intervention and their outcomes in PNG are likely to differ from other nations, being set on a particular trajectory by historical processes that have helped to modify and shape the human ecology, local economy, disease patterns and nutritional status of the different populations there. In particular, the types of cash cropping introduced by colonisers may have a significant impact on health and nutrition, as may urbanisation and industrialisation.

In PNG, the development of gold and copper mining since the 1960s, which has contributed to most of the increase in GNP, was based on the development of gold mining in the early colonial times.

Thus, evaluation of the health and demographic effects of modernisation and intervention require some understanding of the history by which the economic processes currently in operation were put in place. Furthermore, history, culture change and change in behaviour can help in understanding the ways in which economic change is mediated in the production of health, and why some health interventions are more likely to be successful in some populations than others. Behavioural changes associated with modernisation are likely to have an impact on health, but this is difficult to evaluate with existing information in PNG, since the demographic and health indicators commonly used by health planners in PNG do not reveal the complexity of health change and disease experience in the urban setting. Local and regional differences in history, culture and behaviour are as likely to contribute to variation in health outcomes of development and change, as are between-country differences. The wide spread in national-level health and demographic indicators at any level of GNP is probably a function of such variation.

ACKNOWLEDGEMENTS

I thank John Lourie for making available unpublished data from the Ok Tedi Health and Nutrition Project, Papua New Guinea.

REFERENCES

Adels, B. R. and Gajdusek, D. C. (1963). Survey of measles patterns in New Guinea, Micronesia and Australia, with a report of new virgin soil epidemics and the demonstration of susceptible primitive populations by serology. *Am. J. Hyg.*, **77**, 317–43.

Armelagos, G. J. (1990). Health and disease in prehistoric populations in transition. In *Disease in populations in transition. Anthropological and epidemiological perspectives* (ed. A. C. Swedlund and G. J. Armelagos), pp. 127–44. Bergin and Garvey, New York.

Babbot, F. L. and Gordon, J. E. (1954). Modern measles. *Am. J. Med. Sc.*, **228**, 334–61.

Bindon, J. R. and Baker, P. T. (1985). Modernization, migration and obesity among Samoan adults. *Ann. Hum. Biol.*, **12**, 67–76.

Braun, J. von and Kennedy, E. (1986). *Commercialization of subsistence agriculture: income and nutritional effects in developing countries.* Working

Papers on Commercialization of Agriculture and Nutrition No. 1. International Food Policy Research Institute, Washington, D.C.

Briend, A., Hasan, K. H., Aziz, K. M. A. and Hogue, B. A. (1989). Are diarrhoea control programmes likely to reduce child malnutrition? Observations from rural Bangladesh. *Lancet*, **2**, 319–22.

Brincker, J. A. H. (1938). A historical, epidemiological and aetiological study of measles (Morbilli: rubeola). *Proc. Roy. Soc. Med.*, **31**, 807–28.

Caldwell, J. C. (1982). *Theory of fertility decline.* Academic Press, London.

Clements, F. W. (1936). A tuberculosis survey of a Papuan village. *Med. J. Australia*, 253–8.

Darwin, C. (1845). *Journal and researches into the geology and natural history of the various countries visited during the voyage of H.M.S. Beagle round the world.* J. M. Dent and Sons Ltd, London.

Dennett, G. and Connell, J. (1988). Acculturation and health in the highlands of Papua New Guinea. *Current Anthropology*, **29**, 273–99.

Denoon, D. (1989). *Public health in Papua New Guinea. Medical possibility and social constraint, 1884–1984.* Cambridge University Press, Cambridge.

Department of International Economic and Social Affairs (1986). *World population prospects: estimates and projections as assessed in 1984.* United Nations, New York.

Dewey, K. (1979). Commentary—Agricultural development, diet and nutrition. *Ecol. Food Nutr.*, **8**, 265–73.

Dick, G. and McKillop, R. (1976). *A brief history of agricultural extension and education in Papua New Guinea.* Extension Bulletin No. 10. Department of Primary Industry, Port Moresby.

Eggleton, M. (1968). *Report of patrol to Wopkaimin territory.* Olsobip Patrol Report No. 4. Territory of Papua and New Guinea, Port Moresby.

Epstein, T. S. (1972). Indigenous economy. In *Encyclopaedia of Papua and New Guinea* (ed. P. Ryan), pp. 306–14. Melbourne University Press, Melbourne.

Ferro-Luzzi, A., Sette, S., Franklin, M. and James, W. P. T. (1992). A simplified approach of assessing adult chronic energy deficiency. *Eur. J. Clin. Nutr.*, **46**, 173–86.

Finney, B. R. (1972). *Big men and business: entrepreneurship and economic growth in the New Guinea Highlands.* University of Hawaii Press, Honolulu.

Fleuret, P. and Fleuret, A. (1980). Nutrition, consumption and agricultural change. *Human Organisation*, **39**, 250–60.

Food and Agriculture Organisation of the United Nations (1974). *Food balance sheets.* Food and Agriculture Organisation, Rome.

Food and Agriculture Organisation of the United Nations (1980). *Food balance sheets 1975–77.* Average and per capita food supplies 1961–65, average 1967–77. Food and Agriculture Organisation, Rome.

Food and Agriculture Organisation/World Health Organisation of the United Nations (1973). *Energy and protein requirements.* World Health Organisation Technical Report No. 522, Geneva.

Franklin, D. L., Harrell, M. and Parillon, C. (1985). Nutritional functional classification study of Panama. *Food Policy*, **12**, 63–74.

Gibson, R. (1990). *Nutritional Status*. Oxford University Press, Oxford.
Glanville Corney, B. (1913). A note on an epidemic of measles at Rotuma, 1911. *Proc. Roy. Soc. Med.*, **6**, 138–43.
Gross, D. and Underwood, B. A. (1971). Technological change and calorie costs: sisal agriculture in Northeastern Brazil. *Am. Anthropol.*, **73**, 725.
Gunther, J. T. (1972). History of medical services. In *Encyclopaedia of Papua and New Guinea* (ed. P. Ryan), pp. 748–56. Melbourne University Press, Melbourne.
Guyer, J. (1989). From seasonal income to daily diet in a partially commercialized rural economy (Southern Cameroon). In *Seasonal variability in Third World agriculture* (ed. D. E. Sahn), pp. 137–50. The Johns Hopkins University Press, Baltimore.
Hammerskjold Foundation (1976). *What now? Another development*. Dag Hammarskjold Foundation, Uppsala.
Harding (1981). *Urban growth 1980*. National Population Census, Research Monograph No.1. National Statistical Office, Post Moresky.
Harvey, P. W. and Heywood, P. F. (1983*a*). Twenty-five years of dietary change in Simbu Province, Papua New Guinea. *Ecol. Food Nutr.*, **13**, 27–35.
Harvey, P. W. and Heywood, P. F. (1983*b*). *Nutrition and growth in Simbu*. Research Report of the Simbu Land Use Project, Volume IV. Papua New Guinea Institute of Medical Research, Madang.
Hatanaka, S. (1972). *Leadership and socio-economic change in Sinasina, New Guinea Highlands*. New Guinea Research Bulletin No. 45. Australian National University, Canberra.
Hernandez, M. *et al.* (1974). Effect of economic growth on nutrition in a typical community. *Ecol. Food Nutr.*, 283–91.
Heywood, P. F. and Jenkins, C. (1992). Nutrition in Papua New Guinea. In *Human biology in Papua New Guinea. The small cosmos* (ed. R. Attenborough and M. Alpers), pp. 249–67. Oxford University Press, Oxford.
Hitchings, J. (1982). *Agricultural determinants of nutritional status among Kenyan children with model of anthropometric and growth indicators*. Ph.D. dissertation, Stanford University, Palo Alto, California.
Horowitz, I. L. (1972). *Three worlds of development*. Oxford University Press, New York.
Huss-Ashmore, R. A. (1993). Agriculture, modernisation and seasonality. In *Seasonality and human ecology* (ed. S. J. Ulijaszek and S. S. Strickland), pp. 202–19. Cambridge University Press, Cambridge.
Huss-Ashmore, R. and Curry, J. J. (1989). Diet, nutritional status, and economic constraints in rural Swazi households. *Am. J. Phys. Anthropol.*, **78**, 243.
Hyndman, D. C. (1990). The Ok Tedi mining project: problems with the pot of gold. In *Children of Afek. Tradition and change among the Mountain-Ok of central New Guinea* (ed. B. Graig and D. Hyndman), pp. 161–8. University of Sydney Press, Sydney.
James, W. P. T., Ferro-Luzzi, A. and Waterlow, J. C. (1988). Definition of chronic energy deficiency in adults. *Eur. J. Clin. Nutr.*, **42**, 969–81.

Jones, G. J., Willy, D., Lumsden, B., Taufa, T. and Lourie, J. A. (1987). Trace metals in the hair of inhabitants of the Ok Tedi region, Papua New Guinea. *Env. Pollution*, **48**, 101–15.

Kaiser, L. and Dewey, K. (1991). Household economic strategies and food expenditure patterns in rural Mexico: impact on nutritional status of preschool children. *Ecol. Food. Nutr.*, **25**, 147–68.

Kennedy, E. and Coghill, B. (1987). *Income and nutritional effects of the commercialization of agriculture in Southwestern Kenya.* International Food Policy Research Institute Research Report No. 63, Washington, D.C.

Kow, F. (1987). *Assessment of the nutritional status of preschool school children in Papua New Guinea with reference to the appropriateness of international standards.* Ph.D. thesis, University of London.

Lev, L. (1981). *The effect of cash cropping on food consumption adequacy among the Meru of Northern Tanzania.* Working Paper No. 21. Michigan State University, East Lansing, Michigan.

Lourie, J. A. (ed.) (1987). *Ok Tedi health and nutrition project, 1982–1986.* University of Papua New Guinea, Port Moresby.

Lourie, J. A., Taufa, T., Cattani, J. and Anderson, W. (1986). The Ok Tedi Health and Nutrition Project, Papua New Guinea: physique, growth and nutritional status of the Wopkaimin of the Star Mountains. *Ann. Hum. Biol.*, **13**, 517–36.

McGarvey, S. T. (1989). Five-year longitudinal changes in fatness and blood pressure in American Samoa. *Am. J. Phys. Anthropol.*, **78**, 269–70.

McGarvey, S. T. (1992). Economic modernization and human adaptability perspectives. In *Health and lifestyle change* (ed. R. Huss-Ashmore, J. Schall and M. Hediger), pp. 105–13. Museum Applied Science Center for Archaeology Research Papers in Science and Archaeology Volume 9. The University Museum of Archaeology and Anthropology, University of Pennsylvania, Philadelphia.

McGarvey, S. T., Bindon, J. R., Crews, D. E. and Schendel, D. E. (1989). Modernization effects on familial aggregation of Samoan blood pressure. *Med. Anthropol.*, **4**, 321–38.

McGarvey, S. T. and Baker, P. T. (1979). The effects of modernization and migration on Samoan blood pressure. *Hum. Biol.*, **51**, 461–75.

McGrath, G. (1972). *Report of patrol to Wopkaimin territory.* Ningerum Patrol Report No. 16. Territory of Papua and New Guinea, Port Moresby.

MacGregor, W. (1895) British New Guinea. *Proceedings of the Royal Colonial Institute*, **26**, 201–4.

Malcolm, L. A. (1970) *Growth and development in New Guinea—a study of the Bundi people of the Madang district.* Institute of Human Biology Monograph Series No. 1, Madang.

Miller, D. S. (1979). Man's demand for energy. In *Food chains in human nutrition* (ed. K. Blaxter), pp. 23–45. Applied Science Publishers Ltd, Barking, Essex.

National Center for Health Statistics (1977). *NCHS growth curves for children*. Publication No. (PHS) 78–1650. United States Department of Health, Education and Welfare, Hyattsville, Maryland.

National Census. (1981). *Population Census, Papua New Guinea, 1980*. National Statistical Office, Port Moresby.

Norgan, N. G. (1990). Body mass index and body energy stores in developing countries. *Eur. J. Clin. Nutr.*, Suppl. 1, 79–84.

Niemeyer, R., Foeken, D., Okello, W. and Hoorweg, J. (1988). *Nutritional conditions at a settlement scheme in Coast Province*. Ministry of Planning and National Development Report No. 20. African Studies Centre, Nairobi.

O'Dea, K., Akerman, K. and Spargo, R. M. (1980). The effect of transition from traditional to urban lifestyle on the insulin secretory response in Australian Aborigines. *Diabetes Care*, **3**, 31–7.

O'Dea, K., Spargo, R. and Nestel, P. J. (1982) Impact of westernisation on carbohydrate and lipid metabolism in Australian Aborigines. *Diabetologia*, **22**, 148–53.

Organisation for Economic Cooperation and Development (1974). *Development cooperation review*. Organisation for Economic Cooperation and Development, Paris.

Pinstrup-Anderson, P. and Jaramillo, M. (1989). The impact of drought and technological change in rice production on intrayear fluctuations in food consumption: the case of North Arcot, India. In *Seasonal variability in Third World agriculture* (ed. D. E. Sahn), pp. 264–84. The Johns Hopkins University Press, Baltimore.

Policy Planning and Research Division, Papua New Guinea Ministry of Health. (1986). *Handbook on health statistics, Papua New Guinea 1985*. Ministry of Health, Port Moresby.

Policy Planning and Research Division, Papua New Guinea Ministry of Health (1991). *Handbook on health statistics, Papua New Guinea 1990*. Ministry of Health, Port Moresby.

Rabeneck, S. (1982). *The determinants of protein-energy malnutrition among preschool children in Kenya with respect to cash cropping and self-sufficiency in staple food production*. Ph.D. dissertation, Cornell University, Ithaca, New York.

Rowland, M. G. M., Rowland, S. G. J. G. and Cole, T. J. (1988). Impact of infection on the growth of children from 0 to 2 years in an urban West African community. *Am. J. Clin. Nutr.*, **47**, 134–8.

Ruzicka, L. and Kane, P. (1990). Health transition: the course of morbidity and mortality. In *What we know about the health transition: the cultural, social and behavioural determinants of health* (ed. J. Caldwell, S. Findlay, P. Caldwell, G. Santow, W. Cosford, J. Braid and D. Broers-Freeman), pp. 1–26. Health Transition Centre, Australian National University, Canberra.

Ryan, P. (1972). *Encyclopaedia of Papua and New Guinea*. University of Melbourne Press, Melbourne.

Schofield, W. N. (1985). Predicting basal metabolic rate, new standards and review of previous work. *Hum. Nutr.: Clin. Nutr.* **39C**, Suppl. 1, 5–41.

Schuurkamp, G. J. T. (1992). *The epidemiology of malaria and filariasis in the Ok Tedi region of Western Province, Papua New Guinea*. Ph.D. thesis, University of Papua New Guinea.

Schuurkamp, G. J. T., Matango, M., Napil, J. and Vergara, R. V. (1987*a*). Malaria control and the development of the Ok Tedi Mining Project in the Star Mountains of Papua New Guinea: the initial construction phase. *Papua New Guinea Med. J.*, **30**, 27–38.

Schuurkamp, G. J. T., Matango, M., Kereu, R. and Napil, J. (1987*b*). Malaria, splenomegaly and filariasis in the Ok Tedi area of the Star Mountains, Papua New Guinea: three years after residual DDT spray. *Papua New Guinea Med. J.*, **30**, 219–300.

Seymour-Smith, C. (1986). *Dictionary of anthropology*. Macmillan Press Ltd, London.

Shetty, P. S. and James, W. P. T. (1994). *Body mass index. A measure of chronic energy deficiency in adults*. FAO Food and Nutrition Paper No. 56. Food and Agriculture Organisation of the United Nations, Rome.

Smith, A. D. (1994). The politics of culture: ethnicity and nationalism. In *Companion encyclopedia of anthropology* (ed. T. Ingold) pp. 706–33. Routledge, London.

Souter, G. (1961). *The last unknown*. Angus and Robertson, Sydney.

Spencer, T. and Heywood, P. F. (1983). Seasonality, subsistence agriculture and nutrition in a lowlands community of Papua New Guinea. *Ecol. Food Nutr.*, **13**, 221–9.

Swedlund, A. C. and Armelagos, G. J. (1990). Introduction. In *Disease in populations in transition. Anthropological and epidemiological perspectives* (ed. A. C. Swedlund and G. J. Armelagos) pp. 1–15. Bergin and Garvey, New York.

Szathmery, E. J. E. (1990). Diabetes in Amerindian populations: the Dogrib studies. In *Disease in populations in transition* (ed. A. C. Swedlund and G. J. Armelagos), pp. 75–104. Bergin and Garvey, New York.

Taufa, T., Mea, V. and Lourie, J. (1990). A preliminary report on fertility and socio-economic changes in two Papua New Guinea communities. In *Fertility and resources* (ed. J. Landers and V. Reynolds), pp. 35–46. Cambridge University Press, Cambridge.

Tomkins, A. M. (1983). Nutritional cost of protracted diarrhoea in young Gambian children. *Gut*, **24**, A549.

Tomkins, A. M. (1986). Protein-energy malnutrition and risk of infection. *Proc. Nutr. Soc.*, **45**, 289–304.

Tudor, J. (1969). *The handbook of Papua and New Guinea*. Pacific Publications, Sydney.

Ulijaszek, S. J. (1992). Human energetics methods in biological anthropology. *Yearbook of Physical Anthropology*, **35**, 215–42.

Ulijaszek, S. J. (1993). Changes in subsistence practices and diet in Papua New Guinea. In *Food, diet and economic change, past and present* (ed. C. Geissler and D. J. Oddy), pp. 183–206. Leicester University Press, Leicester.

Ulijaszek, S. J. (1994). Persistence and change in food use in Papua New Guinea and its effects on nutrition. In *Nutrition in a sustainable environment*. Proceedings of the 15th International Congress of Nutrition (ed. P. Nestel and M. Wahlquist). Smith-Gordon, London.

Ulijaszek, S. J. and Pumuye, A. (1985). Adequacy of energy and protein intake amongst adult Wopkaimin in the Ok Tedi region. *Papua New Guinea Med. J.*, **28**, 295–301.

Ulijaszek, S. J., Hyndman, D. C., Lourie, J. A. and Pumuye, A. (1987). Mining, modernisation and dietary change among the Wopkaimin of Papua New Guinea. *Ecol. Food Nutr.*, **20**, 148—56.

Ulijaszek, S. J., Lourie, J. A., Taufa, T. and Pumuye, A. (1989). The Ok Tedi Health and Nutrition Project, Papua New Guinea: adult physique of three populations in the North Fly region. *Ann. Hum. Biol.*, **16**, 61–74.

United Nations Children's Fund (1990). *The state of the world's children.* Oxford University Press, Oxford.

Vargas, L. A. (1990). Old and new transitions and nutrition in Mexico. In *Disease in populations in transition. Anthropological and epidemiological perspectives* (ed. A. C. Swedlund and G. J. Armelagos), pp. 145–60. Bergin and Garvey, New York.

Venkatachalam, P. S. (1962). *A study of the diet, nutrition and health of the people of the Chimbu are (New Guinea Highlands*). Papua New Guinea Department of Public Health Monograph No. 4, Port Moresby.

Vines, A. P. (1972). Disease prevalence (morbidity). In *Encyclopaedia of Papua and New Guinea* (ed. P. Ryan), pp. 258–66. Melbourne University Press, Melbourne.

Waterlow, J. C., Buzina, R., Keller, W., Lane, J. M., Nichaman, M. Z. and Tanner, J. M. (1977). The presentation and use of height and weight data for comparing the nutritional status of groups of children under the age of 10 years. *Bull. World Health Organisation*, **55**, 489–98.

Wilson, C. C. (1972). Industrialisation. In *Encyclopaedia of Papua and New Guinea* (ed. P. Ryan), pp. 547–52. University of Melbourne Press, Melbourne.

Wirsing, R. (1985). The health of traditional societies and the effects of acculturation. *Current Anthropology*, **26**, 203–22.

Zemel, B., Worthman, C. and Jenkins, C. (1993). Differences in endocrine status associated with urban–rural patterns of growth and maturation in Bundi (Gende-speaking) adolescents of Papua New Guinea. In *Urban ecology and health in the Third World* (ed. L. M. Schell, M. T. Smith and A. Bilsborough), pp. 38–60. Cambridge University Press, Cambridge.

Index